MW00472045

Praise for *One-Strau*

"Larry Korn virtually brings Masanobu Fukuoka back to life in *One-Straw Revolutionary* by highlighting his experience of more than thirty-five years of study with Mr. Fukuoka. Here we not only get a new look at Mr. Fukuoka's natural farming but also his life in general. For those who have or have not read the insightful *The One-Straw Revolution*, I highly recommend this delightful book about one of the world's great agricultural thinkers."—**JOHN P. REGANOLD**, Regents Professor of Soil Science & Agroecology at Washington State University

"In *One-Straw Revolutionary*, Larry Korn revisits his experiences with Masanobu Fukuoka, one of the most important thinkers in agricultural history. This book is a sort of sequel to Mr. Fukuoka's *The One-Straw Revolution*, clarifying and amplifying that book and then going on to reveal Mr. Korn's own intriguing contributions to the new social and agricultural order."—**GENE LOGSDON**, author of *Gene Everlasting* and *A Sanctuary of Trees*

"*One-Straw Revolutionary* is a profound sharing of the essential philosophy of natural farming translated through the friendship between Larry Korn and Masanobu Fukuoka. Larry's engaging story offers wise insights into authentic practices that honor the community of all life. I deeply resonate with both the author's perspectives and Fukuoka's clear understanding of a revolutionary pathway for creating abundance by honoring the natural patterns of our earth."—**KATRINA BLAIR**, author of *The Wild Wisdom of Weeds*

"I still think *The One-Straw Revolution* is the best book Rodale ever published, and we can thank Larry Korn for bringing it to us. Larry's deep insight into Fukuoka-san's Zen-like approach to farming threw a new light on the organic method of farming and gardening for me, as I was then an editor of *Organic Gardening* magazine. Through Larry, I was able to see that the question is not, 'What can I do next?' but rather, 'What can I stop doing without diminishing the results?' This impulse toward simplicity is the master's great gift to the world, carried forth into the world by Larry Korn."—**JEFF COX**, author of twenty books, including the best-selling *From Vines to Wines* and the James Beard Foundation-nominated *The Organic Cook's Bible*, and former managing editor of *Organic Gardening* magazine

"Larry Korn shines a light on the path that Fukuoka discovered integrating indigenous agriculture with a deep reverence for the land and natural processes. Many revolutions of the sun later, it is clear that the continued illumination of this path is necessary to bring about a stewardship culture of soil, plant, animal, and human. We are fortunate to have a torch bearer in Korn who embodies the words of Taoist sage, Lao T'zu: 'What you do is what you are.'"—DON TIPPING, founder of Seven Seeds Farm and Siskiyou Seeds

"This mind-opening book will provide the proper contextual knowledge and understanding on how nature works for any practitioner involved in farming, ranching, ecosystem restoration, or natural-resource management."—RAY ARCHULETA, conservation agronomist, Natural Resources Conservation Service

One-Straw Revolutionary

One-Straw Revolutionary

The Philosophy and Work of Masanobu Fukuoka

Larry Korn

Chelsea Green Publishing
White River Junction, Vermont

Project Manager: Bill Bokermann
Project Editor: Brianne Goodspeed
Developmental Editor: Makenna Goodman
Copy Editor: Laura Jorstad
Proofreader: Helen Walden
Indexer: Peggy Holloway
Designer: Melissa Jacobson

Printed in the United States of America.
First printing August, 2015.
10 9 8 7 6 5 4 3 2 1 15 16 17 18 19

Our Commitment to Green Publishing
Chelsea Green sees publishing as a tool for cultural change and ecological stewardship. We strive to align our book manufacturing practices with our editorial mission and to reduce the impact of our business enterprise in the environment. We print our books and catalogs on chlorine-free recycled paper, using vegetable-based inks whenever possible. This book may cost slightly more because it was printed on paper that contains recycled fiber, and we hope you'll agree that it's worth it. Chelsea Green is a member of the Green Press Initiative (www.greenpressinitiative.org), a nonprofit coalition of publishers, manufacturers, and authors working to protect the world's endangered forests and conserve natural resources. *One-Straw Revolutionary* was printed on paper supplied by McNaughton & Gunn that contains 100% postconsumer recycled fiber.

Library of Congress Cataloging-in-Publication Data
Korn, Larry, author.
One-straw revolutionary : the philosophy and work of Masanobu Fukuoka / Larry Korn.
pages cm
Other title: Philosophy and work of Masanobu Fukuoka
Includes bibliographical references.
ISBN 978-1-60358-530-9 (pbk.) — ISBN 978-1-60358-531-6 (ebook)
1. Fukuoka, Masanobu. 2. No-tillage. 3. Organic farming. 4. No-tillage—Japan. 5. Organic farming—Japan. I. Title. II. Title: Philosophy and work of Masanobu Fukuoka.
S604.K67 2015
631.5'814—dc23

2015016871

Chelsea Green Publishing
85 North Main Street, Suite 120
White River Junction, VT 05001
(802) 295-6300
www.chelseagreen.com

There is no big or small on the earth,
no fast or slow in the blue sky.

—MASANOBU FUKUOKA

CONTENTS

Author's Notes

THROUGH MOST OF THE TEXT I have referred to Masanobu Fukuoka as Mr. Fukuoka. This English honorific is equivalent to the Japanese *-san*, a term of respect used for both women and men. Mr. Fukuoka's formal title was Fukuoka-sensei. While the word *sensei*, or "teacher," can refer to any teacher, when it is used for teachers possessing superior spiritual understanding the meaning is closer to "sage." Most people addressed him simply as Sensei, which is the way I refer to him in the first-person narrative chapters of this book. When used alone the term reflects both familiarity and respect.

Additional photos of Mr. Fukuoka, his farm, and his travels can be found at www.onestrawrevolution.net.

Introduction

Masanobu Fukuoka (1913–2008) was a Japanese farmer and philosopher from the island of Shikoku. For more than sixty-five years Mr. Fukuoka developed a unique method of farming that has the potential to reverse the degenerative momentum of modern agriculture. Natural farming requires no machinery, no chemicals, no prepared compost, and very little weeding or pruning. Mr. Fukuoka did not plow the soil or flood his rice fields as farmers have for centuries in Asia and around the world. His method requires less energy than any other, needs no fossil fuel, and creates no pollution, yet the fertility of the soil increases with each passing season. Despite this unconventional approach, Mr. Fukuoka's yields were comparable to, or greater than, those of the most productive farms in Japan.

This technique is a demonstration of Mr. Fukuoka's back-to-nature philosophy. "It proceeds from the conviction that if the individual temporarily abandons human will and so allows himself to be guided by nature, nature responds by providing everything."[1] His message shows the way to a brighter future for humanity, a future in which people return to their appropriate place in the world and in so doing find peace within themselves. He considered healing the land and the purification of the human spirit to be one process, and demonstrated a way of life and a way of farming in which that process can take place.

As a young man Mr. Fukuoka left his rural home and traveled to Yokohama to pursue a career as a microbiologist. He was a specialist in plant diseases and worked for several years in a laboratory as an agricultural customs inspector. It was at that time, while he was still a young man of twenty-five, that Mr. Fukuoka experienced a realization that changed his life forever.

He saw that nature was ideally arranged and perfectly abundant just as it was. People, with our limited understanding, try to improve on nature thinking the result will be better for human beings, but adverse side effects invariably appear. Then we take measures to counteract these side effects and even larger problems occur. By now almost everything people are doing is mitigating problems caused by previous misguided actions. All of this activity results only in wasted effort. Mr. Fukuoka believed that people would be better off doing nothing at all from the beginning.

He tried to explain his ideas to his co-workers, but was dismissed as an eccentric woefully behind the times. Finally he decided to quit his job and return to his family farm to test the soundness of his ideas by applying them in his own fields.

For many years Mr. Fukuoka worked to perfect his techniques while living in a small hut in his citrus orchard and had little contact with anyone outside his village. His goal was to give nature free rein by eliminating the hobbling effect of unnecessary agricultural practices. He observed the way plants existed in the wild where great forests grew without help from human beings. He saw healthy rice seedlings growing up in abandoned fields that had not been plowed for many years, and vegetables volunteering in the spaces between the orchard trees.

From then on he stopped plowing and stopped flooding his rice fields. He stopped sowing rice seeds in a starter bed in the spring, and instead scattered them in the autumn when they would naturally have fallen to the ground. Instead of plowing to get rid of weeds he learned to control them by spreading a mulch of rice and barley straw and maintaining a continuous ground cover of white clover. Once he managed to tilt conditions slightly in favor of his crops, Mr. Fukuoka interfered as little as possible with the plants and animals in his fields and orchard. As time went on he found that the less he did, the more productively the land responded. He referred to his method as natural farming.*

---

* Farming as simply as possible within and in cooperation with the natural order, rather than the modern approach of applying increasingly complex techniques to remake nature entirely for the benefit of human beings.

Once he had proven to his own satisfaction that his non-intervention method was superior to scientific agriculture, he began lecturing and writing books and articles about his experience. Mr. Fukuoka had to self-publish his first books because he could not find a publisher willing to take a chance on ideas so far from the mainstream. That changed after the first oil crisis in the early 1970s. Suddenly everyone was looking for an alternative to petroleum-based food production. A publisher finally asked Mr. Fukuoka to write a book introducing his natural farming method and how he came to be farming that way. The result was *Shizen Noho Wara Ippon no Kakumei* (*Natural Farming: The One-Straw Revolution*), which was published in 1975.

The book had little immediate impact in Japan, but when it was translated into English a few years later it became a sensation. People were just beginning to recognize the shortcomings of modern civilization and the environmental damage it was causing. Mr. Fukuoka's message, that people needed do less, rather than more, was like a much-needed balm to a culture that was heading in lockstep down the road of progress with no apparent destination in mind, and many embraced it. "The more people do," he wrote, "the more society develops, the more problems arise. The increasing desolation of nature, the exhaustion of resources, the uneasiness and disintegration of the human spirit, all have been brought about by humanity's trying to accomplish something."[2]

The appeal of *The One-Straw Revolution: An Introduction to Natural Farming*, as the book was titled in English, was not limited to Mr. Fukuoka's philosophy, however. The example of his organic no-tillage farming method was equally provocative. Many farmers and researchers, even in mainstream agriculture, had long known that plowing the soil created a variety of problems. They were trying to develop a no-tillage system for grains and other crops that would avoid using so much energy, burning out the organic matter in the soil, and causing erosion, but no one could figure out how to do it, at least not without drenching the fields with herbicides and chemical fertilizers. Mr. Fukuoka's demonstration showed that it was indeed possible; in fact, he had already been doing it successfully for almost three decades. In the years following the book's publication agricultural scientists from all over the world came to Mr. Fukuoka's farm to verify his work for themselves.

It was these two aspects together, the compelling philosophy *and* the practical application, that made Mr. Fukuoka's message so powerful. Although natural farming contradicts modern society's most fundamental values, the conclusions of science, and traditional know-how that farmers have relied on for centuries, people listened because Mr. Fukuoka spoke with the authority that can only come from knowledge, experience, and tangible results taken together. When visitors marveled at his fields and said they could not believe it was possible to grow crops that way, Mr. Fukuoka modestly said, "Well, the proof is ripening right before your eyes."

The One-Straw Revolution was translated into more than twenty-five languages, Mr. Fukuoka gained an international reputation, and he began traveling abroad. Beginning with his first visit to California, then to Europe, Africa, India, China, and Southeast Asia, he saw firsthand how human activity was turning the earth into a desert. His focus shifted from small-scale, diversified farming to finding ways to rehabilitate the human-caused deserts of the world using the same approach that had been so successful at his own farm in Japan.

Today Mr. Fukuoka is known as one of the founding leaders of the worldwide natural farming movement and one of the great philosophers of modern times. He is respectfully referred to as a sensei in Japan, a *rishi* in India, and a sage in the West, and has won numerous international awards. Despite all of the accolades, however, and despite the lengthy explanations he has given in lectures, books, and articles, few people today would say with confidence that they understand what natural farming actually is. And with all the advantages of his organic no-till farming technique, you would think that many farmers around the world would have followed Mr. Fukuoka's example, but the fact is, few people are practicing natural farming today.

While the reasons for this are complex, natural farming is not. In fact, Mr. Fukuoka considered it "embarrassingly simple." Simple, however, is not a concept that is readily understood or appreciated in today's world. In this book I will attempt to show how simple natural farming actually is, and why doing less, rather than more, is our only real hope for reestablishing a wholesome relationship with the earth.

My first meeting with Mr. Fukuoka occurred one summer afternoon in 1974 when I was twenty-six years old. I had been living in Japan for several years, working on traditional farms and back-to-the-land communes. I often heard people speak of Mr. Fukuoka, always with respect for his spiritual teachings, but no one I met had actually been to his farm or knew the details of his farming method. The stories were incredible, however. One person said that all Mr. Fukuoka did was scatter seeds and then spent the rest of his time meditating. Another had heard, more accurately, that he farmed among the weeds and clover, without plowing, flooding the rice fields, or pruning his orchard trees, but still managed to get high yields. It all sounded fascinating, so one day I decided to visit his farm for myself.

Once I arrived at his village I made my way to the rice fields. It was easy to tell which ones were Mr. Fukuoka's. The rice was shorter than average, the color was dark green, almost olive, and there were many more grains to each head. The surface was covered with straw, white clover, and weeds, and insects and spiders were everywhere. This was in stark contrast with the neighboring fields, which consisted of neat rows of pale rice plants growing in a flooded field with no weeds or insects of any kind. Mr. Fukuoka saw me, came over, and introduced himself. He explained that the reason the rice grew that way was because he did not use any chemicals and the soil had not been plowed for more than twenty-five years.

His orchard was equally remarkable. Instead of the usual rows of neatly pruned trees growing in bare soil, his citrus trees were scattered here and there among a lush ground cover that was waist-high in most places. There were countless other fruit and nut trees, shrubs, berries, and vines, with vegetables growing in the spaces between the orchard trees. Insects buzzed from blossom to blossom and chickens were running all over. At the time, Mr. Fukuoka invited students to work on the farm while he instructed them. I jumped at the opportunity and spent the next two years living in one of the modest mud-walled huts in the orchard. By doing farmwork each day, and in discussions with Mr. Fukuoka and the other student workers, the details of Mr. Fukuoka's method and its underlying philosophy gradually became clear to me.

While I was there, the Japanese edition of *The One-Straw Revolution* was published. After we read it, another student and I decided to translate it into English and try to get it published in the United States. We believed that Mr. Fukuoka's philosophy and agricultural example were too important to languish in Japan, where he had worked for many years in virtual anonymity. When we had what we considered a reasonable draft, I was entrusted to go to the United States to find a publisher.

Shortly after the English-language edition of the book was published a few years later, Mr. Fukuoka received an invitation to come to the United States. I arranged a six-week tour of California, New York, and New England, and accompanied him as translator and guide. It was his first trip outside Japan and his first time flying on an airplane. Seven years later, in 1986, I accompanied him on another six-week tour, which also included Oregon and Washington and appearances at several large international conferences. We met many people along the way, and visited countless farms and a variety of natural areas. It was fascinating for me to see his first impressions when he came to new places and met new people. Every stop was unique and every talk he gave was different.

Although I have not been back to Japan since 1976, I continued to work to make Mr. Fukuoka's way of farming and his philosophy available to people in other parts of the world. I gave talks and workshops and occasionally wrote articles for newspapers and magazines. I also worked as a soil scientist for the California Department of Forestry, at a retail nursery in Berkeley, California, and then at a wholesale nursery in the Monterey Bay area. In 1987 I returned to Berkeley, where I ran my own residential landscaping business for more than twenty years while my daughter, Lia, was growing up. In the spring of 2008 I closed Mu Landscaping and moved to Ashland, a small town in southern Oregon, so I could again work full-time promoting natural farming and helping to build resilient, self-sustaining local communities.

Several of Mr. Fukuoka's books have been translated into English. The first, *The One-Straw Revolution* (Rodale Press, 1978), was meant as an introduction both to his worldview and to the farming methods he developed in accordance with it. In the book he tells the story of how he came to farm the way he did, gives an overview of his philosophy and farming techniques, and discusses his views on such things as diet, economics, politics, science, formal education, and the difficulties humanity created for itself when it chose to separate itself from nature.

In his next book, *The Natural Way of Farming* (Japan Publications, 1985), Mr. Fukuoka gives the details of his farming method and how they evolved over the years. It is mainly practical and is particularly useful to those who are interested in applying Mr. Fukuoka's natural farming methods on their own land. In *Sowing Seeds in the Desert* (Chelsea Green Publishing, 2012), Mr. Fukuoka explains his philosophy in greater detail as well as touching on topics such as evolution, Eastern and Western medicine, the fear of death, the shortcomings of science, economics, genetically modified organisms (GMOs), and the wide-ranging dangers of growing food as if it were an industrial commodity. He also discusses his travels to Africa, India, Southeast Asia, and the United States, and introduces his plan to revegetate the human-caused deserts of the world using natural farming.

Most people only know of natural farming through Mr. Fukuoka's books, but books can take the reader only so far. While he described in detail the form natural farming took given the conditions at his farm on Shikoku Island, many people still hoped he would give an easy-to-follow recipe that could be applied in all locations. In a way he does, because he lays out a path that anyone can use to become reunited with nature. When that happens, people intuitively know what to do in whatever conditions they find themselves. That part of the message often goes unheard, however, because people insist on simple, how-to-do-it solutions that appeal to their analytical way of thinking.

Another hurdle for Western readers is that Mr. Fukuoka explains his philosophy using his own Japanese cultural frame of reference. Terms like *no-mind*, *do-nothing*, and *non-discriminating understanding*, while familiar to Eastern readers, are all but meaningless to most Westerners. The cultural differences go even deeper than the difference between East and West, however. Natural farming is nearly identical to the worldview and the ways of Indigenous people.* They lived within the natural world by using it without depleting nature's ability

* People who originally inhabited a specific territory or region in the world. Until about ten thousand years ago, Indigenous people lived in every corner of the earth, mainly in tribes or tribelets. Each community was distinct and unique, although they shared a similar worldview and code of ethics that allowed them to live sustainably for many thousands of years. The term also refers to the remnants of these cultures that still live today. The word *indigenous* is nearly identical in meaning to the word *native* and, when referring to human communities, is often used interchangeably with the word *tribal*.

to replenish itself. Their culture was informed by the knowledge and practical skills that were passed down in an unbroken chain from the beginning of human existence. Modern society,* on the other hand, has estranged itself from nature and lives in a world of its own intellectual creation. It is based on the assumption that the world can only be understood through analytical thinking and the empiricism of science. This blocks people's access not only to the world as it actually is, but also to an understanding of nature as Mr. Fukuoka experienced and tried so desperately to explain.

I decided to write this book to clear up some of this confusion, and to answer some of the questions that come up time after time about natural farming. I believe it is the first English-language book written *about* Mr. Fukuoka, his work and philosophy, rather than by him. I will try to explain his understanding in a way that may be easier for Westerners to understand because it will be through my eyes—the eyes of a Westerner who grew up in the city and did not have the benefit of an enlightening experience like the one Mr. Fukuoka had before I arrived at his farm.

I have devoted the early chapters of this book to Mr. Fukuoka's story, his philosophy, and natural farming as seen through my eyes and experiences as one of Mr. Fukuoka's students. But those chapters alone cannot explain natural farming and how it fits in with other forms of agriculture. So the latter chapters of the book—chapters 5, 6, and 7— compare natural farming with Indigenous ways, traditional Japanese agriculture, organic agriculture, and permaculture. The ways of Indigenous people and traditional Japanese agriculture have much in common with natural farming. Indeed, Indigenous ways are nearly identical; with Japanese farming there are similarities and also differences. Scientific and organic agriculture, and permaculture, on the other hand, are fundamentally different from natural farming since they are based on the perception and values of modern culture. Over the years, as I have been teaching Westerners like myself about natural farming, I have found that these

---

* Our current worldwide culture that developed within the past eight thousand to ten thousand years, roughly coinciding with the development and rise of agriculture.

comparisons are often the most effective way to explain what natural farming is and is not. And because natural farming is not only about agriculture, it seemed fitting that the final chapter should be devoted to how anyone can use natural farming for personal growth and development, and for leading a richer and more fulfilling life.

I was fortunate to have lived at Mr. Fukuoka's farm and to have been taught by him directly. I worked side by side with him in the rice fields and in the orchard, and traveled with him and his wife, Ayako-san, in the United States. At his request we even visited my parents in Los Angeles for a couple of days. We had a very close personal relationship, although always on a proper teacher–disciple basis. Seeing him at work in the fields, interacting with others, giving lectures, and listening carefully to how he answered questions greatly expanded my understanding of his way of thinking. Not everyone had that opportunity, so in this book I also share my own experiences—what it was like to come to his farm and see his fields for the first time, to live and work with the other students on "the mountain," and to hear how Mr. Fukuoka explained his techniques and ideas in the context of everyday experience.

CHAPTER 1

Mr. Fukuoka's Story

MASANOBU FUKUOKA WAS ON A MISSION. For more than sixty years he toiled at his small farm in Japan to prove that humanity knows nothing, that it should do nothing, and that everything people have done has been wasted effort. It sounds preposterous, but that is the path he took. He was a one-man revolutionary who believed that one straw could change the world. To many he seemed like a quirky country farmer, but by the time he died in 2008 he had countless followers throughout the world who believed wholeheartedly that his vision and example could lead to a better world.

The Journey

Masanobu Fukuoka grew up in a small village on the island of Shikoku, where his ancestral family had lived for hundreds of years. As a youth he worked in the fields and in the citrus orchard of his family farm, describing himself in those days as being carefree and somewhat irresponsible. He began his formal education at the local elementary school, but for middle school and high school he had to travel to Matsuyama city, seventeen miles away. Each morning he rode his bicycle to the train station in Iyo, rode the train to Matsuyama, and then walked the rest of the way on foot—a commute of about an hour and a half in each direction. He was an average student, often exasperating his teachers with his indifference and misbehavior.

Nevertheless, since his father, Kameichi Fukuoka, was the largest landowner in the village and acted as its mayor for many years, Mr. Fukuoka had the opportunity to enroll in the Gifu Agricultural College, near Nagoya. He received a broad education studying rural sociology, English, German, Western philosophy, and ethics. Eventually he settled into the study of modern large-scale agriculture, specializing in plant pathology under the tutelage of Dr. Makoto Hiura, one of the top agricultural scientists in Japan. Science fascinated him, and he soon became one of Dr. Hiura's prized students. After he graduated, Mr. Fukuoka took a job at the Okayama Prefectural Research Station. A year later he began working at the Plant Inspection Division of the Customs Bureau in Yokohama, one of the busiest ports in Japan.

Mr. Fukuoka studied the diseases and insect pests that were found on imported produce and took turns inspecting incoming plants. He enjoyed doing technical research and "was amazed at the world of nature as revealed through the eyepiece of a microscope."[1] After three years there he was hospitalized with acute pneumonia and nearly died. Even after he recovered and returned to work he was still preoccupied with ponderous thoughts about life's purpose and the meaning of life and death.

Then, on the morning of May 15, 1937, Mr. Fukuoka inexplicably experienced a revelation in which he saw "the eternal form of nature." He was dozing at dawn against the trunk of a tree overlooking the harbor. He later wrote, "As the breeze blew up from below the bluff, the morning mist suddenly disappeared. Just at that moment a night heron appeared, gave a sharp cry, and flew away into the distance. I could hear the flapping of its wings. In an instant all my doubts and the gloomy mist of my confusion vanished. Everything I had held in firm conviction, everything upon which I had ordinarily relied was swept away with the wind . . . I felt that this was truly heaven on earth and something called 'true nature' stood revealed."[2]

He tried explaining his vision to his co-workers, that everything was meaningless and that all of humanity's activity resulted in nothing but wasted effort, but he was dismissed as an eccentric standing in the way of progress. At that time people believed science and technology were about to usher in a golden age of abundance and leisure. He decided to leave his job and return to his village home. Rather than trying to explain his understanding using words alone, he would apply it to agriculture, giving it a physical form and so demonstrating its usefulness. He wanted to show that the productive power of nature alone was greater than that of modern agriculture, which relied entirely on human knowledge and technology. After he arrived home Mr. Fukuoka moved into a small hut on the mountainside, where he was entrusted with the care of his father's citrus orchard.

His first attempt at what he thought was natural farming was a complete failure. Mr. Fukuoka's idea was to let nature have free rein so he stopped pruning the orchard trees, which had already been trimmed low and wide so they could be harvested easily. He believed that without further pruning, the trees would revert to their natural forms. As the branches grew out, however, they crossed and became tangled. Within a few years more than four hundred trees withered and died. He realized that once human beings have interfered with the natural pattern, they cannot suddenly desert nature and expect nature to get along on its own. That is not natural farming, but simple abandonment. Once people have affected nature they have a responsibility to repair the damage they have caused. Natural farming is only possible, he realized, when nature is whole, and in most cases that requires a period of rehabilitation.

By the late 1930s the drumbeat of war in Japan became unmistakable. At his father's urging Mr. Fukuoka left the tranquility of his hilltop hut and found a job at the Kochi Prefecture Agricultural Experiment Station on the eastern side of Shikoku. He and his colleagues were expected to increase wartime food production through their research. Mr. Fukuoka also acted as a community extension agent, helping individual farmers increase their yields, and even wrote a "farming tips" column for the local newspaper. But all the while he was carrying out personal research on the side, comparing the yields of crops grown using compost, chemical fertilizers, pesticides, and herbicides with those grown without compost or synthetic chemicals. He found that growing crops using compost and chemicals resulted in marginally higher yields, but not high enough to make up for the cost of achieving them. Once he was satisfied with results of those experiments he never did side-by-side experiments like that again.

As the war was nearing its end in the spring of 1945, even Mr. Fukuoka was drafted into service. He was sent to the front to dig foxholes for the expected invasion of the Allied forces. The land invasion never came, however, and the war ended abruptly just four months later. Grateful to have been spared, Mr. Fukuoka breathed a sigh of relief, tossed away his shovel and his gun, and left for home.

Shortly after the war ended, the Allied occupation forces led by General Douglas MacArthur instituted several important changes. Local officials like Mr. Fukuoka's father were removed from office, and a land reform program redistributed the rice fields more equitably among the villagers. Mr. Fukuoka returned to find that the family's holdings had been reduced to about half an acre of rice paddies and two acres of orchard. Later he was able to purchase eighteen more acres of poor orchard land from his neighbors. He kept half for his natural farm and his son, Masato-san, farmed the rest using organic methods. The rice fields were later increased to almost one and a half acres.

The farmhouse in the village had a courtyard, a few outbuildings, and a small organic vegetable garden outside the kitchen door, but Mr. Fukuoka moved back into a hut in the orchard. He spent the next several years observing the condition of the soil and noting the interaction of the plants and animals living there. He also took long walks into the mountains to discover what he referred to as the natural pattern. Recalling that time, Mr. Fukuoka said, "I simply emptied my mind and tried to absorb what I could from nature."[3]

He noticed that the orchard consisted of very few species—the orchard trees, a few shrubs and perennials, and some scraggly weeds. All of the topsoil had eroded away, leaving the exposed subsoil of hard, red clay. If he had done nothing, nature would have continued in a downward spiral. To repair the damage, Mr. Fukuoka first concentrated on improving the soil and increasing the diversity of species. To loosen the soil he scattered the seeds of deep-rooting vegetables such as dandelion, burdock, dock, and daikon radish. To clean and enrich the soil further he sowed seeds of hardy plants that have substantial, fibrous root systems, including buckwheat, alfalfa, mustard, turnip, amaranth, and yarrow. Then he added a leguminous ground cover of white clover and vetch. The clover enriches the soil and is effective at suppressing weeds. The vetch thrives in the winter when the clover does not grow as readily.

He also planted a variety of trees and shrubs including nitrogen-fixing acacia trees to improve the deeper layers of the soil. The acacia trees grew quickly, so after eight or nine years he would cut them down and use them for firewood or as a building material, leaving the roots to decay over time. As he removed the trees, he planted others in different places in the orchard so the cycle of soil-building continued. Eventually the soil became rich and the structure of the orchard came to resemble that of a natural woodland. By the time I came to the farm about thirty years after he began this process, the soil was fertile and there were more than thirty different kinds of fruit- and nut-bearing trees, berries of all kinds, and vegetables growing everywhere. There were also chickens and ducks, a few goats, some rabbits, and beehives.

Mr. Fukuoka believed that if nature were given the opportunity, it would respond by providing everything, so he was always looking for ways to minimize his meddling. As he wrote in *The One-Straw Revolution*:

> *The usual way to go about developing a method is to ask, "How about trying this?" or "How about trying that?" This is modern agriculture and it only results in making the farmer busier.*
>
> *My way was just the opposite. I was aiming at a pleasant, natural way of farming which results in making the work easier instead of harder. "How about* not *doing this?" "How about* not *doing that?"—that was my way of thinking. I ultimately reached the conclusion that there was no need to plow, no need to apply fertilizer, no need to make compost, no need to use insecticide. When you get right down to it, there are few agricultural practices that are really necessary.*[4]

When he returned to the orchard Mr. Fukuoka found that the natural systems were so badly damaged that he had to do many tasks himself that later became unnecessary. Once the permanent ground cover of soil-building plants became established, for example, he no longer needed to apply fertilizer. Once the diversity of plants was restored it provided habitat for a wide range of insects, so he no longer needed to make and apply organic insecticides. Eventually he was hardly doing anything, just sowing seeds, spreading straw, cutting back the ground cover from time to time, and waiting for the harvest.

He applied the same simplicity in his grain fields, where he grew a crop of rice and one of barley in the same field each year. In the fall he sowed the seeds of winter barley and white clover directly into the ripening stalks of rice. By the time the rice was harvested the young barley plants were already covering the surface of the soil. Then he spread the uncut rice straw back over the field. The clover, the straw mulch, and the fact that there was no gap between the two field crops made it difficult for weeds to get a foothold. Then he sowed the seeds of the following year's rice crop directly into young barley plants.

By growing his grains in succession this way he avoided the common tasks of plowing, growing rice seedlings in a starter bed in the spring and then transplanting, weeding, fertilizing, and irrigating the crops. Rather than the tidy, controlled appearance of his neighbors' fields, Mr. Fukuoka's displayed the wild exuberance of natural growth. His soil improved each season and soon reverted to the character of soils in natural grasslands. Despite their unkempt appearance, the yields of rice in Mr. Fukuoka's fields equaled and often surpassed the yields of his neighbors who used chemicals and the most advanced technology available.

During the years Mr. Fukuoka was developing his farming method he mainly lived in the orchard, but this did not bring him the serenity he had hoped for. He described himself then as being bad-tempered and disagreeable even to his own family.* From his mountain retreat he witnessed with indignation the degeneration of both the land and Japanese society. The Japanese single-mindedly followed the American model of economic and industrial development. Farming became a business and food a commodity. As human and animal labor were replaced by machines and agricultural chemicals, the population shifted from the countryside to the growing industrial centers. Farmers came to care less about the quality of the food they produced than they did about how much they could get for it. With these changes came pollution and spiritual disintegration. Mr. Fukuoka felt frustrated that he could not do more to stop this process despite having been shown how humanity could live more harmoniously in the world. He was given

* Mr. Fukuoka and his wife, Ayako-san, raised five children together, four girls and one boy.

a gift of insight, and that, he felt, brought with it the responsibility to help others. It was a burden he carried with him for his entire adult life.

By the late 1960s Mr. Fukuoka felt a new sense of urgency. He decided that he could not, in good conscience, remain silent any longer. He became more involved, presenting his thoughts in open forums, writing books and articles, and appearing on radio and television. He invited students to live in his orchard so he could pass his practical farming knowledge and experience on to them, and invited anyone who was interested, including scientists and government officials, to visit his farm and see the abundance in his fields for themselves. Still, very few people followed his example.

That all changed with the publication of the Japanese edition of *The One-Straw Revolution*, in 1975, and its subsequent English-language translation in 1978. Mr. Fukuoka finally had an opportunity to visit the world outside Japan, and what he saw alarmed him. He was shocked by California's nearly treeless plains of withered annual grasses, what others proudly called California's golden hills. He understood that California's Mediterranean climate lacked the summer rains of Japan, but he believed that people were playing a major role in turning the Golden State into a desert with industrial methods of agriculture, poor water management, overlogging, and overgrazing. After visiting other parts of the country he came to refer to this as America's ecological disaster. What he saw in India and Africa gave him an idea of the magnitude of the worldwide ecological crisis. From that time on he devoted all his energy to solving the problem of reversing desertification using natural farming methods.

Mr. Fukuoka believed that most of the deserts of the world were caused by human activity, and that efforts to rehabilitate them were only making things worse. He believed the deserts could be revegetated using broad-scale aerial seeding. He advocated encasing the seeds of as many species as possible in clay pellets that also contained microorganisms. If the seeds of all types of plants were made available, nature would find the most appropriate course of action given present conditions. He referred to this as the Second Genesis. Most important, people's preconceived ideas and assumptions would be left out of the decision-making process.

In 1985 he traveled to Somalia with hundreds of pounds of seeds to test his theory. Most of the seeds were sent to him by Japanese homemakers who answered his call to save the seeds of fruits and vegetables as they were preparing meals. He also brought several hundred fruit tree seedlings. The plan could not be carried out, however, due to interference by the Somali government, so he moved on to a refugee camp in a remote part of Ethiopia where a group of Japanese volunteers were already providing assistance. There, at least, he was able to show the refugees how to plant trees and care for them, and how to grow their own vegetable gardens.

In 1987 Mr. Fukuoka made the first of his five trips to India, where he was received as a *rishi*. By the time he arrived, *The One-Straw Revolution* had been widely distributed and read, and had already been translated into four local dialects. Almost half of the population of India still lived by subsistence farming, so his low-cost, spiritually based form of agriculture was readily appreciated and adopted. Several farming communities had already changed to his no-tillage, non-intervention method by adapting it to their local crops and conditions.

He spoke of natural farming as being akin to Mahatma Gandhi's way, "a methodless method, acting with a non-winning, non-opposing state of mind,"[5] and praised those who followed it. At Visva Bharati, the university founded by Noble laureate Rabindranath Tagore, he received the school's highest honorary degree from a former prime minister of India, Rajiv Gandhi. In a later visit he had an hour-long conversation with the then prime minister, P. V. Narasimha Rao. That meeting was widely covered on television and in the newspapers.

On October 9, 1997, Mr. Fukuoka had a particularly memorable visit with natural farmer Bhaskar Save at his fourteen-acre orchard/farm located in South Gujarat just north of Mumbai (formerly Bombay). There the main cash crops, coconuts and sapota,* occupy about ten acres in a diverse natural orchard, a coconut nursery occupies two acres, and another two acres are used for growing grains, vegetables, and other seasonal crops for home consumption.

Mr. Save began farming in 1953 when chemical agriculture was just being introduced to India. He became a model recruit for the Green Revolution. After a few years he noticed that he kept doing more and

---

* *Manilkara zapota.*

more, and spending more and more, but earning less. He also saw that the condition of his soil and the vigor of the plants were deteriorating. This caused him to question the value of scientific agriculture and look for an alternative. He found it in the undisturbed forests near his farm where trees, shrubs, and ground cover plants grew abundantly without tilling, fertilizing, weeding, or any other human assistance. Mr. Save came to the conclusion that the fundamental blunder of agricultural scientists was their attempt to increase productivity when nature was already most abundant just as it was. Beginning in 1960 he began applying these insights at his orchard farm.

By the time Mr. Fukuoka visited, Mr. Save's farm was among the most productive in India, far more productive than the neighboring farms that used modern scientific techniques. It was covered with a diversity of vegetation, yet Mr. Save, his family, and the few women who helped with harvesting hardly had to do anything to maintain it. The farm grew in fertility and retained much more water as it gradually came to resemble its natural condition.

Although he had no formal agricultural education, Mr. Save remarked that he didn't need formal training because "my university is my farm." Mr. Fukuoka was amazed as he toured the orchard farm while sitting in a bullock-drawn cart with Mr. Save at his side. As they passed from one lush section of the orchard to the next, Mr. Fukuoka kept saying, "Wonderful! Wonderful!" When the group reassembled after the tour and someone asked him what he thought of the farm, Mr. Fukuoka said, "I have seen many farms all over the world. This is the best. It is even better than my own farm!" Today India has more practical demonstrations of natural farming—and more interest in it—than anywhere else in the world.

Perception

Most people see natural farming primarily as an agricultural method, but the farming is only a physical demonstration of Mr. Fukuoka's view of the world. The foundation of natural farming is what Mr. Fukuoka saw that morning in Yokohama. This provides the framework for everything that followed.

He saw for the first time that existence is an undivided unity in which everything is interconnected, ideally arranged, and teeming with life force. Time exists as an ever-changing continuum of the present moment with the past and the future embedded within it. As humans we are an essential part of this unity but are generally blocked from experiencing it that way because we consider ourselves a separate entity. When we give up the idea that we are separate, true nature reveals itself and we are free to take our proper place within it. We are no longer a part of nature as seen from a distance, we *are* nature.

Mr. Fukuoka and other Asian spiritualists see the world as constantly unfolding and changing, a viewpoint that is different from what we are used to in the West. David Hinton, a translator of ancient Chinese poetry and other classical Chinese texts, explains this in an interview with *The Sun* magazine:

> *We think of time in linear terms, whereas in ancient China they thought of existence as a burgeoning forth, an ongoing generative present in which things appear and disappear in the process of change. And this constant birthing goes on both in the physical world and in human consciousness, for consciousness is as much a part of that process as surf or a rainstorm or blossoms opening in an almond orchard . . . existence is alive . . . Things perpetually move and change, appear and disappear. Clouds drift. Wind rustles wildflowers and trees. Day fades into night, and night into day. Seasons come and go, one after the other. You die. Other people are born. On and on it goes. Everything is moving all the time without pause, without beginning or end . . . We resist that change here in the West. We want permanence, an immortal soul that allows us to escape death . . .*[6]

In the West we believe that there is some permanent identity inside of us. This sense of self is most closely associated with our mental process—our rational, analytical faculties. That is summed up in Descartes' celebrated "I think, therefore I am," sometimes translated as "I am thinking, therefore I exist." But it is precisely this assumption that alienates us from the world.

Asian spiritualists believe that the minute you think of yourself as a distinct and permanent entity you remove yourself from the world of constant change. When they referred to the natural mind, they generally meant an empty consciousness that perceives the world with mirror-like clarity, without the constant interpretation of our inner dialogue. When the world is seen with that kind of clarity, it becomes possible to see *yourself* in every flower and blade of grass. "An empty mind that mirrors the world puts the world inside of us."[7]

For Mr. Fukuoka there was one reality: the world exactly as it exists without intellectual distinctions or judgments of any kind. People's minds divide phenomena into dualities such as life and death, yin and yang, joy and sorrow. Only when people do that do these things come into existence. In nature, they do not exist. Understanding nature, how it works, or why any of this exists when there could just as easily be nothing at all lies beyond the reach of human intelligence. There is no need to trouble ourselves trying to figure out the purpose or the precise meaning of life. Instead we should just accept our gift, and do so with gratitude.

So far, this is basic Asian spirituality. Becoming one with the whole is the goal of virtually all of the Japanese arts and Asian religions. Each has its own program or structure to help people achieve this goal, but the aim is always a higher consciousness in which the individual merges with existence itself. The "self" disappears, so the spirit can act through you without resistance. Some of these programs are quite rigorous and take years to complete. Often, students never achieve the ultimate goal. Sometimes they become so comfortable within the program's structure that they decide the practice itself is sufficient. Regardless, all of these practices have a final stage in which the program is no longer necessary.

Two things set Mr. Fukuoka's natural farming apart from the other disciplines. The first is that there is no program and no structure. Simply serving nature by living humbly and providing for your own daily needs is seen as the most direct path toward self-awareness. There is no meditation, yoga, required reading, or anything else. It is a methodless method. As the land gets closer and closer to its original form, the mind of the farmer also finds its way back to its original state. You become free and able to simply enjoy life.

You do not have to study Asian spirituality to experience life in this way. It is accessible to anyone at any time whether you are a farmer or not. Mr. Fukuoka felt that programs were unnecessary because in nature programs do not exist. They only seem valuable because they help to correct the unnatural conditions created by people's separation from the natural world. Why clutter things up with another overlay? If a teaching program is used, students have to heal not only their alienation from nature, but also their attachment to the program itself. The goal of natural farming is to return to the heart of nature. The most direct path toward achieving this goal is to clear the mind of preconceptions and attachments, and just live here and now.

Mr. Fukuoka's approach also provides for your daily necessities and restores damaged landscapes. The study of Noh drama or flower arrangement as a spiritual art form ultimately leads to the creation of fully realized beings who will, presumably, do good works in the world, but their physical manifestations are of little practical value. With natural farming, the end products are food, shelter, and a clean, wholesome environment. It produces both realized beings *and* a healthier and more abundant world.

Natural farming sees the world as a seamless, indivisible whole. This perception arises without conscious effort when the world is experienced without intellectual interpretation. Mr. Fukuoka referred to this as non-discriminating awareness. Conventional perception, or discriminating understanding, sees the world as a collection of discrete parts. The individual is an observer, and the world is that which is being observed. It is a willful, intellectual process, which results in a fragmented and incomplete perspective.

Here's another way to look at it. The world is completely interconnected, flowing and unfolding within the present moment. It has no intrinsic qualities, it simply is. People, in their attempt to organize experience into a logical framework, break up this unity by seeing the trees as distinct from shrubs, insects, minerals, and other things. These elements are further subdivided into types of trees, shrubs, insects, and minerals, and they are all given names. Then concepts such as up and down, east and west, large and small, fast and slow, and self and other are added. Finally, values such as strong and weak, good and bad, beautiful and ugly, and primitive and civilized are also included. Before you know it, people have created an entirely new reality known only to them. In nature, none of these things exists. This is the way people have set themselves apart from and in opposition to the natural world. Today almost everyone relies on intellectual understanding to interpret and get along in the world. Before modern culture arose, no one did.

When people see something, they automatically interpret it according to previous experiences. Someone sees an oak tree and thinks, *Oak tree: provides shade and habitat for wildlife, produces acorns and firewood. Oak trees are good.* Then they store that information in their "oak tree file." The next time they see an oak tree they do not see the tree itself, but only the information they have on file for it. The tree itself is none of those things, of course, it just *is*. When someone sees a ladybug they immediately think *beneficial insect*. When they see an aphid, the thought *pest* automatically pops up. What they are seeing is not the oak tree, the ladybug, or the aphid itself, but the *idea* of those things, along with the values they have learned to associate with them. Those values are most often based on our perceived notions of how the things benefit or harm human beings.

Our arm's-length relationship with nature is plainly revealed in language. We "observe" nature, are "surrounded by nature," or take a few days off to "reconnect with nature." In his seminal work *The Unsettling of America*, Wendell Berry took issue with the Sierra Club's statement of purpose, which reads in part: ". . . to explore, enjoy, and protect the nation's scenic resources . . ." His issue was with the word *scenic*. "A scene," he wrote, "is a place 'as seen by the viewer.' It is a 'view.' The appreciator of a place perceived as scenic is merely its observer, by implication both different and distant or detached from it."[8] It is as if the world were merely a backdrop for the human drama.

While I was standing at the rim of Crater Lake this past summer I happened to overhear the following exchange between a couple who were standing next to me. "Now, *that's* what I call a world-class view!" the woman said. "Yes, it *is* beautiful, but it's not as good as Yosemite," her companion replied. First the lake becomes a "world-class view," then it is ranked against other views using an arbitrary human standard. That is understandable, I suppose, because the scenic view and the judgment exist in the world in which the couple live; the *reality* of Crater Lake and Yosemite do not.

Words are useful in the world of discrimination because they define a reality that has been previously agreed upon. That reality is taught in the home, in schools, at church, in the media, and in literature. Words are not adequate, however, for describing the world as seen without discrimination. Words originate in the intellect and are, therefore, limited. Even the ethereal poetry of William Blake and Walt Whitman can only lead us to the brink of transcendental understanding; from there we must complete the journey on our own.

Although Mr. Fukuoka was quite fond of talking and wrote many books and articles, he ultimately felt that his words were useless. He was particularly frustrated when he could not find words to adequately communicate his awareness of nature. He noticed that when he used the word *nature* each person saw only their idea of nature, not its reality. Eventually he came to the conclusion that there was no need to give it a name or try to describe it. When someone asked him to describe nature as he understood it, he would take them on a walk through his orchard and not say a word.

The ones who see the world most clearly and with no distinctions are infants and small children. They hear the songs of birds and feel the warmth of the sun directly and without judgment. As children get older, however, they are introduced to the world of relativity. This begins with language and continues with formal education and perhaps religious education as well. Mr. Fukuoka explains,

> *When a child first becomes aware of the moon, that child is simply filled with wonder. Then after a period of time the child learns to discriminate between a subject, "I," and an object, "the moon." The child comes to know the thing called the moon as "other." So even in the structure of human language, human beings are taught to set themselves apart from nature. The intimate and harmonious relationship between people and nature that once existed—which we can see in the children's instinctive wonderment—becomes cold and distant.*[9]

Children intuitively know their place within nature, but as adults we need to work to find our way back, should we choose to do so. The challenge is to return to our original mind, the awareness we had before we became aware of ourselves. As people, we can never *actually* separate ourselves from nature or from the limits of the biological world, but when we no longer feel spiritually connected with nature, we come to believe that the natural law no longer applies to us. That, then, gives people the freedom to act in the world in any way they like, and to do so with impunity.

The Role of Science

Science is the most obvious way people manifest discriminating understanding in the physical world. It studies nature as an objective, mechanical entity through observation, empirical analysis, and then reconstruction. That is, it breaks reality into pieces, studies each element as an independent unit, and then attempts to put the pieces back together.* The assumption is that science can, and eventually will, come to understand how nature works. For some scientists it is simply a matter of curiosity, but for the most part the role of science in today's society is to find ways to manipulate nature for the benefit of human beings.

The world is completely interconnected, however, so studying its parts by "isolating them from other variables," as scientists do in their experiments, is not helpful for understanding the whole. An object as seen in isolation from its natural context is not the real thing anymore. When it comes to the natural sciences, a research method that takes all of the relevant factors into account is not possible, so when the conclusions of such research are applied in the real world things always seem to go wrong, and often in unexpected ways. We just can't think of everything. Most scientists are aware of this, but they conveniently sweep it under the rug hoping the things they could not measure, observe, or comprehend will not become too big a problem when their findings are amplified by technology and released into the world.

In *Sowing Seeds in the Desert*, Mr. Fukuoka explains the futility of the scientific approach:

> *If knowledge of the whole (one) is broken into two and explained, and then these are divided into three and four and analyzed, we are no closer to understanding the whole*

---

* The root of the word *science* comes from the Proto-Indo-European *skei* and the Greek *skhizein*, to cut, split, divide, separate, and later the Latin *scientia*, knowledge.

> *than we were before. When we do this, however, we come under the illusion that knowledge has increased. But can we say that by endlessly repeating our divisions and analyses and then gathering up all the fragments, we have advanced human knowledge in any meaningful way? . . . Furthermore, it throws us into confusion.*[10]

That is to say, it leads us into a world of endless questioning. Our craving for information becomes insatiable whether that knowledge has any practical value or not. The more information we generate, the more distant and fearful the world becomes.

It is true that science and technology have created many things that we consider useful, even indispensable, but in most cases these things only seem to be useful because society has created unnatural conditions that make them *appear* to be so. It begins when people alter the environment thinking it will stimulate nature to serve them more efficiently. This leads to unexpected consequences, and science is called upon to save the day. The solution, however, is usually a hastily conceived, stopgap measure that slows the immediate damage but does not correct the source of the problem. This leads to second- and third-generation problems, which are more damaging and even harder to repair. From then on we become dependent on the "solutions," which never would have been needed if we had left things alone in the first place.

Here are some examples from agriculture. When a field is plowed, the natural vegetation, which is habitat for insects, is destroyed. When the crops are attacked by insects there are no predators nearby to keep them under control. Pesticides are used to limit the losses and are considered to be helpful. Eventually, they become indispensable. If the habitat for a diversity of insects had not been eliminated in the first place, the natural insect balance would have kept the damage to a minimum without need for further intervention. Plus, the pesticides kill bees and other pollinators and their toxic residues find their way into the soil, leading to the next generation of problems.

When the fertility of the soil is depleted through plowing, people need to resupply nutrients for subsequent crops. Chemical fertilizer is applied and is thought to be beneficial, then it becomes essential. If the soil had not been plowed and a permanent ground cover of soil-building plants had been grown instead, the fertility of the soil would have been maintained without the need for chemicals. The fertilizer also changes the pH of the soil and causes pollution, sowing the seeds of future difficulties.

It's the same with herbicides. When the soil is plowed, weed seeds from deep in the soil are stirred up and given an opportunity to sprout. Tillage also creates bare ground, which is a magnet for fast-growing weeds. Herbicides are used and are thought to be a lifesaver. Ironically, plowing is often done specifically to eliminate weeds. Actually, it *causes* them to return, putting the farmer on the "plow–weed" treadmill.

This process occurs in human society as well. Formal education seems to be valuable because it teaches skills that help the student get along in the human-created world. If that student were living in a culture where people's lives were directly connected to the natural world, formal education would not only be useless, it would hinder the student's ability to learn from the world intuitively. When a child studies classical music, for example, that child becomes insensitive to the sounds of nature and no longer recognizes them as the standard for beauty. The child comes to think of beautiful music as whatever their particular culture had agreed upon as being beautiful.

> *Teaching music to children is as unnecessary as pruning orchard trees. A child's ear catches the music. The murmuring of a stream, the sound of frogs croaking by the riverbank, the rustling of leaves in the forest, all these natural sounds are music—true music. But when a variety of disturbing noises enters and confuses the ear, the child's pure, direct appreciation of music degenerates. If left to continue along that path, the child will be unable to hear the call of a bird or the sound of the wind as songs . . . The child who is raised with an ear pure and clear may not be able to play the popular tunes on the violin or the piano, but I do not think this has anything to do with the ability to hear true music or to sing. It is when the heart is filled with song that the child can be said to be musically gifted.*[11]

Mr. Fukuoka uses nature, without distinctions or interpretation, as his fixed standard. In every discussion, whether about medicine, diet, education, agriculture, the arts, politics, economics, or anything else, his standard is always the same. Modern society has no such standard. We gave that up when we alienated ourselves from nature and chose to live in the relative world of our own ideas. It became impossible to decide on the absolute value of anything and difficult to know how to live. Most people just pick something to believe in and make their decisions using that as their standard. It could be anything—nationalism, existentialism, hedonism, science, libertarianism, or some religious doctrine. Often the decision is based on values learned in the home or at school, or on perceived self-interest. With so many competing viewpoints and no way of evaluating them with certainty, the world becomes an extremely confusing place.

Natural farming works best when the landscape is as close as possible to natural conditions. In most cases that involves at least some rehabilitation. People gain the greatest personal benefit from practicing natural farming when their internal landscape is as closely aligned to the natural order as possible. That, too, usually involves a period of readjustment. For most of us, that process begins by unlearning most of the things we were taught when we were young. It is a process of letting go, of "whittling away" unnecessary thoughts and preconceptions. Finally, when we are left with "nothing," the world becomes peaceful and welcoming again. When I first heard Mr. Fukuoka say, "There is no need to understand the world, just enjoy it," I felt like a huge weight had been lifted from my shoulders.

CHAPTER 2

My Travels in Japan

WHEN I SET OUT FOR ASIA on the SS *President Cleveland* in 1970 I had no idea that I would end up living for almost four years in the countryside of Japan. I did not have a fixed itinerary so I suppose anything was possible, but what actually happened, the sequence of events that led me to Masanobu Fukuoka's farm and my lifelong involvement with plants, soil, and natural farming, was something I never could have predicted or imagined.

For some reason I have always been interested in Asia. In college at Berkeley I studied history, specializing in the history of China. After I graduated, I decided to travel to Asia to see what it was like there. I had no particular plan other than taking a passenger ship from San Francisco to Yokohama, then working my way to Southeast Asia, India, or wherever. I had about a thousand dollars, a backpack with what I guessed I would need, and a visa for Japan. I was twenty-three years old and had never traveled outside the United States before.

A few of my friends saw me off to streamers and shouts of "bon voyage!" as the ship pulled away from the pier. It was exhilarating to sail under the Golden Gate and into the sunset. I was finally going to Asia.

Large passenger ships are all but gone today, but at the time they were the least expensive way to travel to Asia. People used them like buses. I met many interesting people on the ship. One was a young lady named Kazuko who was returning to Japan after visiting family in Stockton. I also met a missionary couple who were on their way to the

Philippines, a journalist from Hong Kong, an Asian scholar heading to Taipei, who I later learned was an agent for the CIA, and two World War II veterans who were retired and living in Guam. Later I would visit them all. I traveled sixteenth-class (out of sixteen), which turned out to be a dormitory next to the engine room. After the first few stuffy, noisy nights I decided to sleep on the deck under the stars.

Kazuko invited me to stay with her and her grandparents at their home north of Tokyo. Then we spent a month or so traveling in the snowy northern prefectures, eventually making our way to Kyoto, the ancient capital and cultural center of Japan. We had met a New Yorker named Dominic on the ship who was traveling to visit friends there, so we decided to pay him a visit. He introduced us to a wonderfully eclectic community of Japanese and Westerners who welcomed us as if we were family.

For Westerners, Japan is a place where you either feel a strong affinity for the culture right away and stay for a while, or you have difficulty adjusting to the crowds, the fast pace, and the rigid social customs. Those people leave within the first six months, the initial length of a typical tourist visa. Most of the Westerners in this community had lived in Japan for at least several years and everyone was studying one of the Japanese arts, such as pottery, *shakuhachi* (bamboo flute), Noh drama, religious studies, martial arts, calligraphy, or textiles.

The Japanese were from all walks of life, but generally they were freethinking young people who had dropped out of the mainstream. Dropping out, even for a short time, was a very serious decision for them because it meant that they could never really get back in. They were considered slackers. Some of them were members of a loose-knit group of poets, wanderers, and idealists who called themselves *buzoku*, or tribal people. They had previously referred to themselves as "the Bum Academy," but when their interest turned to returning to the land and living as Indigenous people had lived in ancient times, they thought the new name was more appropriate. They were closely connected to the back-to-the-land movement in the United States, although there were far fewer of them in Japan. They established a network of communes in some of the most remote and beautiful places in Japan and invited me to visit, giving me directions to the communes and to people who would let me stay with them along the way. Kazuko needed to return home, so I saw her off at the train station (not an easy good-bye) and then headed for the open road.

Over the next two or three months I hitchhiked to communes in the Japanese Alps and the islands of Shikoku and Kyushu before making my way to Kagoshima city to catch the supply boat to the crown jewel, a commune on Suwanose Island known as the Banyan Ashram.

Suwanose, the Burning Island

The small diesel ship that sailed to Suwanose, the *Toshima* (ten islands) *maru*, had an irregular schedule because it could only sail and unload its cargo in calm or relatively calm seas. When I arrived in Kagoshima I learned that the next sailing would be in about a week, so I found an inexpensive hotel near the wharf and set out to see the town.

Kagoshima is one of the most beautiful cities I have ever been to. The city itself is on the shores of Kagoshima Bay. Sakura-jima, island of cherry blossoms, is a conical volcano that rises in the middle of the bay. It happened to be erupting intermittently while I was there . . . *and* the cherry trees were in bloom. I took a ferry to the island several times to explore it, and spent the rest of my time wandering through the city or along the water's edge, not doing much of anything, really.

The trip from Kagoshima to Suwanose was rough and not at all pleasant for me, so I was glad when we arrived around one o'clock in the afternoon on the day following our departure. The pier was so small and the coast so rocky that the *Toshima maru* had to anchor outside the reef while a small fishing boat ferried passengers and cargo back and forth to the island. Then, loaded with supplies on our backs, we made our way up the steep path to the village of eight or ten houses, then through the bamboo forest to the ashram. I had no idea then that when I left the island just a few months later I would be a completely different person.

Suwanose is one of a string of small islands known as the Ten Islands Chain. These islands are just north of the Ryukyu Islands, which extend south past Okinawa and almost as far as Taiwan. These were the islands Paleolithic seafarers used as stepping-stones thousands of years ago to migrate to the main islands of Japan. Suwanose is an active volcanic island that rises steeply out of the sea to twenty-six hundred feet. Mount

Ontake, the volcano in the center of the island, was erupting more or less all the time while I was there, sometimes for a day or two, sometimes for a week or more at a time. It is the most active volcano in Japan and remains one of the most active in the world. Typically there would be a loud BOOM that shook the ground followed by a giant cloud of gray ash rising to fifteen thousand feet or more. When the wind brought the cloud overhead, ash rained so heavily that everyone had to go inside. Sometimes the ash rain lasted for several days.

The center of the island is mountainous and uninhabited, but there is a broad plateau that is arable. Streams flow down from the mountain mainly during the frequent rainstorms. The island is densely covered with shrubs, bamboo, and pine forests, but villagers and ashram members managed to clear small plots for growing sweet potatoes, squash, melons, and a few vegetables. The villagers also grazed cattle and goats. A warm current passes the island so the sea is filled with many kinds of fish, shellfish, and turtles. The diet for both the villagers and the ashram members consisted of rice, miso soup, sweet potatoes, soybeans, tofu, fish, and occasionally bananas. Once while I was there a cow fell off a cliff to its death, so we had beef for a few days.

Suwanose was probably settled and abandoned countless times over the years depending on the activity of the volcano. The earliest recorded history of the island goes back to 1813, when the largest eruption on record occurred. Two women, who were seven years old at the time, told the story of how they fled to neighboring islands: "As soon as the eruption began there were great rumblings, and a rain of fire stones which burnt the houses. People took refuge in Nanatsu-ana (Seven Hole Cave) at the eastern end of the island and stayed there for several days. As the rain of ash and fire kept falling the people decided to escape to Nakano-shima and Akuseki Islands. At Kirishi Beach they found their boats buried beneath volcanic ash and had to dig them out with hoes before we could escape."[1]

The island remained uninhabited until May 1883, when people from nearby Amami Oshima came to settle. They built a village and cleared fields for sweet potatoes. The present islanders are direct descendants of these settlers. Just six months later the volcano erupted again, wiping out all the crops the settlers had planted and causing a famine. According to one account, "The islanders wandered in the mountains and along the sea in search of anything edible. As the fire of the blazing volcano illuminated the sky, the search for food went on night as well as day. Some died, many fell sick from malnutrition and poisoning when in desperation the starving settlers turned to eating unknown weeds and berries."[2]

In 1971, when I first visited the island, there were about forty villagers and about ten or twelve people living at the ashram. The ashram, named after a beautiful banyan tree near the village, was founded by Nanao Sakaki, Gary Snyder, Sansei Yamao, and several other members of the *buzoku* in 1967. A few thatched-roof structures had already been built, all from materials collected from the nearby forests, as well as a kitchen and a few sheds. There was no electricity or modern conveniences. The conditions were primitive, in keeping with the mission of the tribe.

The ashram referred to itself as a karma yoga ashram, a Hindu ideal in which personal realization is achieved through selfless service. For me, at least at the beginning, that just meant a lot of very, very hard work for which I was woefully unprepared. I was confident in my skills for getting along in the modern world—after all, I'd graduated from college with reasonably good grades, figured out how to avoid the draft, traveled to Asia when I was just twenty-three years old, and was doing fine. But my experience in the "wild" was limited to a few family camping trips when I was a child. I didn't know how to work, use or take care of tools, do carpentry, or plant or care for a vegetable garden, and my cooking skills were minimal at best. I felt completely helpless.

Like most new arrivals I was assigned to clearing fields of bamboo so they could be used for growing sweet potatoes. The work was very hard and my body was not conditioned. My thoughts were scattered and moving much too quickly for the pace of the island. I did not want to be working in the fields. I wanted to be climbing on rocks at the seashore, hiking to the volcano, or back in the hut reading a book . . . *anywhere* but where I was. There was little private time at the ashram. I was deprived of my customary pleasures and the things I normally relied on to relieve stress. That made me anxious and eventually angry, not toward the others, but with the world in general. The first few times I took my turn cooking, the food was awful and it was ready two hours late. I felt like I was letting everyone down no matter what I did. I became withdrawn and moody. In short, I was a mess.

The others must have noticed what I was going through. They did not blame me, but they did not try to cheer me up, either. They had probably seen this happen to other city people when they first came to the island. Nevertheless, I stuck it out and dragged my aching body to the fields every morning after breakfast.

Then, one afternoon, everything changed. I was working alone that day. I remember noticing that the pick was swinging into the soil just as before, the roots were coming loose, but it was effortless, as if someone else was doing the work and I was standing by watching. I was actually enjoying the sensations of my back muscles stretching out with each stroke and reaching down to pull the roots free. In fact, I was loving it. I became one with the bamboo roots and felt like I knew them personally. Then I looked up and saw the island in all its beauty as if for the first time, the trees swaying in the wind, the birds, the open blue sky. At that moment there was no place I would have rather been than right there in the field digging those roots. When I heard the cow bell announcing that dinner was ready I didn't want to leave.

I had become present, centered on what I was doing and nothing else. The Buddhist term for that is *mindfulness*. My mind became clear and my spirit joyful. Work had become play and my confidence returned. I had learned how to work.

The others noticed the difference right away and started treating me differently. One Japanese woman smiled and said, "It looks like you had a good day in the fields today." Someone else showed me how to sharpen the sickle and the pick with a whetstone and file, and another took me into the kitchen and gave me tips on how to prepare food more efficiently and more quickly. One fellow, who had been there for more than a year and had barely said two words to me before, told me the story of what he went through when he first came to the island. It was a story remarkably similar to my own. I had finally been welcomed into the community. I saw how the change in my attitude had given others a lift, and that was helpful to the group as a whole. I realized that working on my own personal awareness was not a selfish pursuit, it was actually an effective way to make real and positive changes in the world.

A few days later Naga-san, one of the founding members of the ashram, mentioned that a few of the men were going to the other side of the island for a few days and he asked if I wanted to come along. Oh my gosh, yes! We started early the next morning and headed straight up the mountain, stopping at the crater for lunch. The volcano was resting that day, but still the roar was deafening. The crater itself is about five or six hundred feet deep and almost a mile wide. Steam was spewing from vents along the sides and the bottom of the caldera. Looking out you could see other islands as faint green dots in the blue sea. Then we continued on, making the trip in about seven hours. We stayed in a cave that was just like the one I imagined villagers had huddled in during the eruption of 1813, fished and foraged for wild edible plants, and generally enjoyed ourselves.

For the next several months I continued to dig bamboo roots, gather firewood from the woodlands and cow dung from the pasture, and plant sweet potatoes. I also helped construct a small house. That involved carrying logs down the mountain for posts and beams, weaving bamboo lattices for the walls, and making a roof of thatched bamboo. I also helped the villagers with some projects in town and went fishing beyond the reef once or twice, but my main interest and responsibility continued to be farming.

One evening I walked to the fields after dinner. The soil, which was almost pure volcanic ash, sparkled in the moonlight. The top of the volcano was glowing an eerie red, getting brighter and then dimmer as if the mountain were breathing. Ontake-san was not in full voice that night, it was only rumbling, kind of humming to itself. I looked out at the half-planted field of sweet potatoes and could hear the rustling of the bamboo in the wind. I stood quietly for a moment and took it all in. Then the soil spoke to me. From then until now, everything I have done has involved plants and soil in one way or another, a gift I could never have expected or imagined.

It had been just over four months, my extended visa was running out, and I was starting to hear the call of the open road, so I left Suwanose Island and my community of new friends and traveled south, always taking the slowest and least expensive method of transportation. I stopped on several other small islands on my way to Okinawa, Taiwan, and finally Hong Kong. I wasn't ready to leave the Pacific islands quite yet, so instead of continuing on to India I decided to treat myself to a vacation (within my extended vacation) and headed east to the Philippines and then Micronesia. After five or six months there, and more adventures, I returned to California where I enrolled in the soil science and plant nutrition program at UC Berkeley.

Back to School

It was lucky that I happened to study soils at Cal instead of one of the other university agricultural programs that had been taken over by agribusiness. The students and professors at Berkeley were there for only one reason—we all loved the soil. There was research, of course, but most of it addressed basic questions about the nature of the soil itself, not just how to use the soil for profit.

When I graduated from high school I remember breathing a sigh of relief because I would never, ever, have to take another science course again. Now, ironically, I found myself committed to a year of basic chemistry, physics, and biology, as well as courses in botany, soil chemistry, soil microbiology, plant pathology, plant nutrition, plant physiology, and a host of other lab and field courses. It was different, though. This time I was applying science to something I was really interested in. I didn't care much for the laboratory work, but I loved being in the field, learning about landforms, about how soil develops over time, and about the relationships among plants, soil, and microorganisms as they exist in natural landscapes.

The best college course I ever took was Soils 105. It was a six-week summer course designed mainly for people who would one day map soil, need to read a soil survey, manage resources for a government agency, or otherwise do soils work in the field. There were about twenty of us, mainly soils students from Berkeley and UC Davis, but there were a few forestry students as well. We traveled all over California and parts of western Nevada, stopping at specific sites to dig holes with an auger or analyze road cuts that revealed the soil profile. Then we worked in teams to describe the soil as someone would if they were doing a soil survey. We looked at the climate, the topography, the color, texture, structure, and other physical properties of the soil, and the relationship between various soil types and the vegetation. We camped most of the time, but sometimes stayed in a university dorm or town hall.

Of all the many lectures I attended during that time, one stands out in my mind. The professor, who usually radiated nothing but tranquility and good vibrations from the front of the classroom, was strangely dour as he shuffled his notes before he began. "Today," he said, "we are going to talk about agriculture." He proceeded to tell us what happens when the soil is plowed: The vegetation is eliminated; the soil structure is broken down, making it hard for air and water to circulate freely; the various layers of the soil are mixed together or inverted; the communities of microorganisms are sent into chaos; the long strands of fungi that are so important for feeding the plants and keeping them healthy are chopped apart; the organic matter is burned at an accelerated rate, reducing the fertility of the soil and its ability to retain water; and the bare ground is left open to erosion. Then he went on to discuss the effects of applying chemical fertilizer.

For a year and a half we had been learning about the magical world of naturally occurring soil, where minerals, microorganisms, root hairs, earthworms, and such all existed together in a blissful state of harmonious perfection. The soil had become our special friend. Then someone comes along and ruins it all by dragging a plow through it. How outrageous! Some of the students, including me, were nearly in tears. Others were angry. Then one young woman near the front raised her hand and asked, "If plowing is so bad, why do we do it?" The professor replied, "Because we don't know any other way to grow enough food." I filed that bit of information and remembered it later when I happened upon Mr. Fukuoka's farm.

Shortly before I graduated I received a letter from my friend Bill, whom I'd met in Kyoto shortly after I came to Japan. He said that he and his wife, Hiroko, their young son, Taichi, and a few others had left the city and were living on a farm in the mountains north of Kyoto with about an acre of land. They had decided to become farmers. I had just spent two years sitting in lecture halls and working tediously in the labs, so it didn't take long for me to decide to head back to Japan and join them.

Shuzan Valley

The Shuzan Valley, where Bill and Hiroko lived, is located about thirty miles north of Kyoto. It takes about an hour to drive there from the city on a narrow, winding road through cedars, pines, and a mixture of deciduous trees, mainly oak, walnut, and maple. Azaleas grew in the understory. The first time I saw the valley it reminded me of Shangri-la.

Except for the occasional car and the power lines, being in Shuzan was like going back in time, reentering Japan's traditional Tokugawa period of the 1700s. There were the rice fields and the vegetable gardens, the thatched-roof farmhouses, the bamboo groves, and the neatly trimmed Japanese cedar trees growing on the mountainsides. The irrigation system was the same as in the past, and it still operated perfectly. Women worked in the fields in traditional blue-and-white cotton clothing, and the air had a scented mountain freshness that was both exhilarating and intoxicating. I had to pinch myself occasionally to be sure I wasn't dreaming.

It was 1974, just twenty-nine years after the end of the Second World War. On the surface it was hard to see, but many profound changes had taken place since then. The konki, a sort of oversized rototiller, came to the valley in the early 1950s, replacing draft horses and oxen. Chemical fertilizers, herbicides, and pesticides were introduced shortly after that. Thanks to these technologies, the need for human labor was dramatically reduced. It took only an hour or so to spread herbicide over a one-acre field, but weeding the same field by hand would have taken three or four people several days to complete. Using synthetic fertilizers rendered the back-breaking job of making compost unnecessary, although the villagers still maintained compost piles for their kitchen gardens.

The new methods were promoted by the Allied occupation forces and were backed by the Japanese agricultural cooperatives, so almost everyone made the switch to scientific farming right away. It was not as apparent then as it is now that the modern methods created pollution, depleted the soil, and left the crops weaker and more vulnerable to insects and diseases.

All of those labor-saving devices cost money, so most of the young people had to move to the city to find paying jobs. The apparent savings in labor were replaced by labor in urban factories. Between 1955 and 1970 the rural population of Japan dropped from 75 percent to 10 percent. Today it is less than half that. The farmers who remained were either elderly or the families of the oldest son who remained to carry on the family's ownership.

This mirrored what occurred in the United States during the same period, but there were also some differences. Except for a few broad plains, the agricultural areas in Japan are narrow valleys and the size of individual fields is quite small. The agricultural land could not easily be joined into mega-farms as happened in the United States to accommodate larger and larger machinery. Also, in the United States, when farmers left the farm, because they either defaulted on loans or moved to the city for personal reasons, they lost ownership. In Japan, even after the rest of the family moved to the city, the eldest son remained, providing continuity of ownership. Relatives would return for visits and maintain a connection to their rural tradition. Even today many city Japanese know how to forage for wild edibles and how to prepare simple medicines from wild plants.

Bill and Hiroko's farmhouse was where the village headman had lived. It was probably built in the mid-1600s, around the same time the daimyo's* castle was built on the hillside overlooking the valley. It was similar to the other farmhouses, with its dramatic thatched roof and outbuildings, but it was larger than the others, had an oversized central room that was used for meetings, and had two ornamental gardens, which we rescued from a tangle of brambles and weeds.

It was also the only house with a large, white-walled storage building. Once, when the owner of the property came to visit, he opened the storage building and let us look inside. There were suits of armor, old furniture, and musty piles of books and papers. The owner did not come often, thank goodness. He had the idea that our community was a Buddhist school and we were his pupils. We gave him the nickname Captain Zen because when he visited, he expected us to sit meditation for two hours in the morning before we could go out to work in the fields. Of course we showed proper deference. We were in Japan, after all, and he was our benefactor.

The house had been abandoned for more than ten years and the fields were an eyesore so we went to work and fixed everything up. The neighbors watched us, tentatively at first. We must have been

---

* Daimyo were feudal lords who ruled over specific regions of Japan until 1867. They were subordinate to the shogun, who wielded absolute authority, although the shogun was technically appointed by the emperor to rule in his name.

quite a sight with our beards, long hair, overalls, and tie-dyed shirts, but they saw that through our work the house was restored and the fields were again tidy and filled with crops. Pretty soon they started dropping by for tea, probably more out of curiosity than anything else at first. Gradually the visits became more regular and more relaxed, and eventually we were accepted into the community. We were invited to take part in cooperative farming activities like planting and harvesting rice, and clearing weeds from the irrigation channels. The villagers were desperate for labor and we were more than happy to do our part.

Our idea was to grow crops the same way the Japanese had grown them in traditional times. We had a relatively large vegetable garden in front of the house and about an acre of irrigated rice fields. There was a limitless supply of manure from a nearby farmer who kept a few cows confined in his barn. He welcomed us to put on our tall rubber boots and muck out the stalls to our hearts' content. There was also an endless supply of rice hulls from the local mill. We dug the manure and rice hulls into the soil with a rototiller and made compost for top-dressing the vegetables. The soil improved noticeably in a very short time.

We frequently asked our neighbors for advice and they were more than happy to oblige. Most of them did not understand why we were so interested in the old way of farming, but with some, we clearly struck a chord. One winter we grew vetch* as a cover crop, just as they had before the war. The neighbor's fields were bare during the winter so ours really stood out with their bright purple blossoms. A few of us were standing next to one of the fields one day when an elderly fellow pedaled by on his bicycle. With tears in his eyes he thanked us for growing that particular cover crop. He told us that it had been more than twenty-five years since he had seen a field like ours and it brought back happy memories.

Another time I was working in the garden in front of the house when an elderly woman, who was permanently stooped as a result of a lifetime of working in the fields, called to me from the street below. "Don't forget to mulch those eggplants," she said. "They like their roots

---

* *Vicia villosa.*

to be cool in this hot weather." I bowed and thanked her, then I went to the barn, got an armful of rice straw, and laid it beneath the plants.

Word of our farm got out and soon friends from Kyoto started to visit. We also became a stop on the itinerary of travelers who had set out from Europe and trekked through the Middle East, India, Nepal, Southeast Asia, and Hong Kong, finally ending up in Japan. The visitors from Kyoto came mainly on weekends. Some enjoyed the novelty of working with us in the fields, but we did not expect them to work. We were glad to provide a haven where they could hike in the mountains, swim in the river, or just relax and enjoy the gardens. The world travelers who stayed for more than a couple of days, however, were expected to join us in whatever work we were doing in the fields at the time and help in the kitchen.

Life was good. Hiroko gave birth to two children at the farm, both with the aid of the local midwife, Ono-san. The first, my godson Raison, arrived at night during a heavy snowstorm. When we called Ono-san to tell her we needed her help she said she would come right over—on her bicycle! It's true that bicycles are the traditional means of transportation for midwives in Japan, but it was cold and snowing and Ono-san was more than seventy-five years old, so we decided to pick her up in the truck instead. When the second child, Honami, was born, I had already returned to the United States, but I heard that Ono-san was on hand for her delivery, too.

One winter everyone took off for a few months and I was left alone to take care of the farmhouse. I spent most of my time sitting by the stove studying Japanese characters and reading books about Asia. I also visited neighbors and sat by *their* stoves, eating mandarin oranges and talking about life in the valley. A frequent topic of conversation was how things had changed over the past twenty-five years. Most of the changes came as a direct result of mechanization and the resulting sudden depopulation. The remaining villagers were simply too old and too few to keep up with the tasks necessary to ensure the land remained as healthy and productive as it had been. This neglect was easy to see in the abandoned houses and fields that dotted the landscape, especially in the more remote areas, but the place where the damage was most severe was not apparent to an average visitor. It was in the forests and woodlands.

In traditional times, branches were trimmed, trees were coppiced, the undergrowth was cut back from time to time, and some trees were removed to discourage overcrowding. Since those jobs had been neglected the woodlands became overgrown, with evergreens gradually replacing the sun-loving deciduous trees. Evergreen trees create a dark and relatively sterile understory, so diversity decreased, as did the forest's ability to store water and control flooding. Bamboo groves got out of hand because people no longer harvested the shoots for food or the stalks for various other uses. Wild boars and monkeys migrated from higher elevations and became bothersome in the rice fields and in the vegetable gardens.

Slopes surrounding the rice paddies were also neglected, endangering the animals that had become dependent on the annual cycle of human-caused disturbances. The Oriental stork, fireflies, and butterflies, species that the Japanese have wistfully associated with the traditional rural way of life, were disappearing. Hawks that had depended on farmers' mowing the strips around the paddies so they could see their prey were becoming more and more rare.

For hundreds of years Japanese farmers created an agricultural landscape that followed a predictable pattern. Plants and other animals adapted to the altered environment and came to rely on it. When people stopped doing the things that were needed to maintain that landscape, all species suffered.

Harvesting Igusa

I had first heard about Masanobu Fukuoka while I was on Suwanose. A few visitors who came to Shuzan also mentioned him, but no one I talked to had actually been to his farm, so he and his farming remained a mystery. One day I decided to travel to Shikoku, visit his farm, and see for myself. It was late spring and we had already finished transplanting the rice. I walked along the narrow paths near the river until I reached the small town a few miles away and boarded the bus for Kyoto. After visiting friends there for a couple of days and wandering through some temple gardens, I continued south.*

---

* There are more than sixteen hundred Buddhist temples in Kyoto. Most have gardens open to the public.

I was running low on funds so I stopped in Okayama Prefecture where a friend suggested I could find work harvesting igusa,* a tall rush used for weaving the top surface of tatami mats. This plant is only grown in a few places in Japan, but a fellow who came to Shuzan had grown up in one of them. It is one of the few seasonal jobs where farmworkers travel to help with the harvest. He gave me some contacts, and then added with a smile, "You know, harvesting igusa is the hardest agricultural job there is in Japan."

"Great," I said, "bring it on."

My friend had called ahead so Ueda-san, the igusa farmer who was to be my boss, was expecting me when I arrived. He explained what would be expected and asked if I was prepared to work that hard. I assured him that I was. "Fine," he said and showed me where I would stay in one of the two rooms set aside for the ten or twelve of us who would be there for the harvest. Then he added, "We're getting up at three o'clock tomorrow morning."

Growing igusa requires exceptional skill on the part of the farmer. Starters must be grown for a little more than a year before they are transplanted to flooded fields in late November or early December. This is a difficult and uncomfortable job since the weather is very cold then and the temperature of the water is near freezing. By the end of July the plants grow to more than four feet tall. The harvest occurs during the hottest, most humid time of the year. The stalks are cut using a straight-bladed sickle at ground level, one handful at a time. The harvester then shakes the reeds to get rid of the short or misshapen ones and piles them nearby. Another worker bundles them into sheaves. The bundles are dipped into a liquid bath of clay and laid out to dry for a day or so. This helps protect the stalks and preserves their aroma and light-greenish color. If rain comes during the drying period, even for a short time, the crop is ruined.

We arrived at the fields just as it was beginning to get light. Ueda-san stood in the bed of the truck looking west for any signs of thunderstorms. He had listened to the weather report, but trusted his experience more. Then he gave the word and we began to harvest. It was, indeed,

---

* *Juncus effusus.*

hard work, but once I got the hang of it I was able to keep up with the others all day long. A group of villagers gathered at the edge of the field looking at us and talking together. Actually, they were looking at me. They had never seen a *gaijin** working in the fields before, let alone harvesting igusa. I smiled and waved and continued working.

We finished when it was too dark to see anymore and returned to the farmhouse covered in mud. Then we took a bath, had dinner, and went to bed at about eleven o'clock. During dinner one of the other workers said to me, "You know what those villagers were talking about, don't you? They were guessing how long you will last."

The next morning, and for the next twelve mornings, we continued with the same routine until the harvest was complete. That evening another farmer from the village asked if I would like to help him harvest his crop, but I politely told him I had other things I needed to do. The next morning I was back on the road with a few hundred dollars in my pocket and the satisfaction of knowing I'd made it all the way to the end.

* A non-Japanese person.

CHAPTER 3

The Natural Farm of Masanobu Fukuoka

I ARRIVED AT MR. FUKUOKA'S VILLAGE a few days later and made my way to the rice fields. It was easy to see which fields were his because they were filled with clover, straw, and weeds; insects were everywhere. Mr. Fukuoka happened to be working nearby and came to greet me. He was a short, slender man, dressed in the boots and work clothes of the average Japanese farmer, but his white wispy beard, sparkling eyes, and alert self-assured manner gave him the appearance of a most unusual person.

We introduced ourselves and chatted for a few minutes. I told him about my experience at the ashram on Suwanose Island and about our farm in the Shuzan Valley. "So," he said, "you are familiar with growing rice. Have you ever seen rice like this before?" I told him that I had not. "That's because these fields have not been plowed for more than twenty-five years." A jolt ran through me. Not plowed for more than twenty-five years? Really? I flashed back to that soils class where the professor said that it would be great if we didn't have to plow the soil, but we simply did not know how to grow food any other way. This was the example no one knew about! I felt like I had just discovered the Holy Grail of organic no-till agriculture.

I had heard that Mr. Fukuoka allowed students to live and work on his farm, so I asked him if I could stay for a while to "receive his teaching," a rather formal way to put it, but it acknowledged that I understood what our relationship would be. He agreed, then added, "The way I farm here is a little different from what you are used to so it would be good if you kept an open mind." He told me how to find the path to the orchard where I would meet the other students; they would show me around.

Looking back, I can see that everything I had done in Japan until then had prepared me for what was to follow. I could speak Japanese reasonably well, especially the farming terminology, and had experience living communally in a rustic setting without modern conveniences. I knew how to use Japanese farm tools and do most of the basic farming tasks, and I could work a full day without complaining. I also knew how important this long-term example of organic no-tillage grain growing would be to the worldwide agricultural community. But the thing that prepared me most for understanding Mr. Fukuoka's philosophy was getting on that ship in San Francisco with few possessions and no plan, only the certainty that everything would work out well.

Living in the Orchard

The small village where Mr. Fukuoka grew up is about seventeen miles from the city of Matsuyama. The Fukuoka family probably settled there centuries ago. The current farmhouse was built by Mr. Fukuoka's grandfather and is typical of the others in the village. The farm consisted of about one acre of rice and barley fields, and a ten-acre citrus orchard situated on a hill overlooking the Inland Sea.* It took about twenty minutes to walk from the fields to the orchard.

I made my way up the narrow, winding road that led to "the mountain," where I was greeted warmly by several young Japanese who were just returning from work. Hide-san showed me around. We walked through a ground cover of mustard, radish, buckwheat, lupine, and weeds that was more than waist-high in most places. Hide-san had to hack at vines and branches occasionally to clear the paths, which were no more than a foot wide and only existed where people or animals had traveled recently. He showed me the spring and the cistern, the

* The small sea amid the islands of Honshu, Kyushu, and Shikoku.

pond, his favorite fruit trees and berry bushes, the shiitake logs stacked neatly beneath some trees, the chicken coop, the toolshed, and a few places where you could look out over the valley and see all the way to Matsuyama Bay. There were mulberry trees, horse chestnuts, ginkgo, sumac, bamboo, and *so* many plants I had never seen before all growing together with no apparent order or design. Each hillside, terrace, and meadow was different from the last.

At one point I asked Hide-san where the vegetable garden was. He stopped and thought to himself for a moment, then looked at me and said, "You mean you haven't noticed yet? The entire orchard is a vegetable garden." He picked off the flowers of some mustard plants, then the leaves of three or four leafy plants that I didn't immediately recognize. "Care for a salad?" he asked with a smile. Then he asked me to follow him a few steps off the trail to what seemed like an overgrown patch of weeds. He parted the plants to expose a few cucumbers, sweet potatoes, and a squash. From then until the end of the tour he kept pointing as we walked. "There's a patch of potatoes here, some melons and a patch of raspberries over there, burdock, tomatoes, and onions there." There were vegetables growing everywhere, not to mention the abundance of wild edible plants. "You'll get to know where everything is in about a month or so," he said. Finally we came to the small hut that would be my home for the next two years. Yu-san was my "hut mate" along with one or two others as the need arose.

Our one-room shelter also served as the group kitchen. The entryway had a dirt floor but most of the living area was a raised platform with tatami mats. Bedding and personal belongings were neatly stacked in one corner. The kitchen supplies, including bulk beans and grains, a sack of potatoes and another of sweet potatoes, jars of flour, pickles, miso, and dried fruit, were in another. There were also utensils, pots, woks, and kettles blackened from years of use. Bundles of dried herbs were suspended on strings inside while other vegetables and fruit hung to dry outside under the eaves. A large earthenware water jar with a dipper was near the front door.

Meals were cooked over a wood-burning fireplace in the center of the room. The various pots and kettles hung in turn over the fire from an adjustable hook that was suspended from the rafters. Smoke from the fire left the inside of the hut through two holes on either side of the ceiling. That worked reasonably well most of the time, but still, the ceiling was coated with a fine layer of soot. The first night I was there we had a delicious meal of steamed rice with barley, sautéed vegetables, and miso soup. After dinner we stayed up late talking under the stars.

The next morning a rooster began crowing before dawn. Soon after, the other workers emerged from their huts. One tended to the chickens and ducks and let them out of the coop. Another went to the spring with buckets to replenish the water jar. The designated cook for the day, Yu-san, lit the fire and prepared a breakfast of porridge made from the previous evening's leftovers, miso soup, rice, a raw egg, and pickles. As we ate, we planned the day's work, what tools we would need, and who would be responsible for what. They were in the middle of cutting the orchard ground cover, a monthlong task, so on this particular morning there wasn't very much to discuss.

Hide-san showed me how to sharpen the long-handled scythe on a whetstone and demonstrated how to use it. "Start at the bottom of the hill and work to the top," he said. "Use long sweeping strokes to cut the ground cover in open areas and quick pulling strokes, like this, when you need to be more precise, like around the trunks of the trees or near patches of vegetables. Just leave the trimmings on the ground. Now you try it." I had used the tool before, but no one had ever instructed me so my technique was clumsy at first. Hide-san made a few suggestions, then said, "Don't worry, you'll be getting lots of practice. We still have more than half the orchard to do." He showed me how to scrape root borers out of the tree trunks and seal the wounds, and how to thin the fruit on each tree so the branches would not get so heavy that they might break off. Then he left to do another job somewhere else in the orchard and I was left alone.

Until then I had been so focused on getting my bearings and meeting everyone that I didn't have time to just look around appreciate where I was. Finally I could relax for a few minutes and let it all soak in—the puffy clouds, the green leaves, the blue sky, the flowers, the fragrances; it was so peaceful and everything seemed to be in perfect order. I sampled a few mandarin oranges, ate some turnip flowers and a handful of blueberries, then picked up the scythe and went back to work.

At the end of that first day's work we decided to go for a swim. The pond was created to catch runoff from the rains and the springs in the area. The water was used to irrigate the rice fields in the valley below. The setting among the trees and the bamboo was as picturesque as you could imagine. There were fish, tadpoles, frogs, water spiders, dragonflies, and turtles. Egrets and herons also showed up from time to time. Cattails grew in the shallow water along the shore. There was no one else living in the mountains so we felt free to swim naked if we felt like it.

On the way back we stopped at a group of three or four persimmon trees with fruit that was just ripening. The best ones were at the top so the others climbed up the trees to get them. It was more my style to stay on the ground and throw grapefruits or poke at them with a bamboo pole. That technique was surprisingly effective, especially after Yu-san showed me how to fashion the end of the pole into a kind of hook. As we sat gorging ourselves beneath the trees, Enta-san, one of the other workers, said that pretty soon the tangerines, persimmons, pomegranates, figs, apples, pears, grapes, and melons would all be ripening at the same time so we should be careful not to eat too much fruit at once. There was silence for a moment, then we all burst out laughing. How could anyone manage to find a problem with such abundance? Even Enta-san saw the humor in his comment and we all facetiously agreed to try to control ourselves.

When I arrived at the farm there was a core group of five residents. Hide-san had been there the longest and had the most comprehensive understanding of the farmwork, so when Sensei was not around he was informally recognized as the group's leader. His plan was to stay for one more year, then start his own natural farm. Yu-san was quite a presence with his stocky build, shaved head, and booming voice. He had lived as a Buddhist monk for several years before deciding that life outside the meditation hall was more to his liking. Enta-san was quiet and very pleasant to be around. He cared for the chickens, calling them to the coop every evening and rewarding them with a few seeds. Kin-chan (literally "golden boy") was young and just learning to be a farmer. He had gone to a missionary school and knew the Bible extremely well. He became a Buddhist for a while, but lately had given up on that, too. A few months later we were joined by Tsune-san, fresh from a one-year tour of organic farms in the United States. It was a very pleasant, hardworking community with a warm sense of camaraderie. Maybe it's because it has been so long, but I cannot remember any serious disagreements.

There were no modern conveniences on the mountain. We lived almost entirely on what we grew ourselves and the resources of the area. Sensei had his students live in this semi-primitive way because he believed it helped us develop the sensitivity and sense of place necessary to practice natural farming. Living this way was part of the teaching. He gave us thirty dollars each month for things that were impractical to produce on a small scale, like cooking oil and soy sauce. He did not pay his students for working there, but no one complained. We were living in one of the most beautiful places imaginable, all of our needs were taken care of, and we were being instructed by a man whom we believed had great wisdom. All of this was freely given and we felt that it was more than adequate compensation.*

There was no spirit of competition. We never raced one another to see who could harvest a swath of barley the fastest, or anything like that. We were more interested in how *well* we did the job than how quickly. Some were more experienced at using the tools and doing farmwork than others and some had physical limitations so the output of each worker varied, but that was not important. What *was* important was that you gave your best effort. Ideally, you did the same job each day a little better than you had done it the day before. Many of the jobs were repetitive and went on for days or weeks. The tedium was a challenge for some, but I thrived on it. It was meditative and, in a way, cleansing. There was nothing to think about day after day except what was right in front of you.

New people, mostly from the city, showed up from time to time not knowing what to expect. We welcomed everyone and included them in the fieldwork, the chores, and the late-afternoon dips in the pond, but we did not baby them. It was similar to the treatment I received when I'd first arrived at the ashram on Suwanose Island. If they could keep up,

---

* Mr. Fukuoka instructed students at his farm for a period of about fifteen years. When students were not living in the orchard he occasionally hired local villagers to help him, especially during the harvest. He paid these workers competitive wages.

great, they were assimilated into the community. If not, they eventually decided for themselves that it was time to move on. We wished them well as they headed back down the mountain and back into the world. While I was there many people passed through for a week or ten days but few were interested in agriculture, which was too bad since Sensei was so willing to teach anyone who was willing to learn.

Vegetables, Clover, and Weeds

After a while the days began to blend with one another. The sun made its familiar pass across the sky, clouds came and went, and the moon went through its phases again and again. The jobs changed with the seasons but the daily routine did not change much at all. Sometimes Sensei asked us to postpone what we had planned because he had something else for us that day, but for the most part we knew what had to be done and were on our own.

He often worked with us, at least for part of the day. He was a tireless worker. Even at more than sixty years of age he bounded up and down the orchard hillsides like a goat. We all had trouble keeping up with him. When he worked, it was like watching someone doing a graceful dance. We always treated him with deference, of course, but when he was part of the work crew he laughed and told humorous stories right along with everyone else. If he saw what he considered careless work, however, his laughter stopped rather abruptly. Another thing that riled him was when he saw someone mishandling or mistreating tools. He was stern at times, but never overbearing.

One morning Sensei appeared carrying a sack of seeds just as we were finishing breakfast. He asked us to follow him to a clearing that had a vigorous growth of mustard, daikon, clover, and weeds. "Today we are going to sow vegetables," he announced. "I began preparing this area last fall by mowing the weeds before they set seed, and then sowing the plants you see here to enrich the soil. I did that just as the summer weeds were dying back but the winter weeds hadn't sprouted yet.

"Each time you mow, the weeds become weaker and the clover becomes stronger. Just leave the cuttings on the ground. They will hide the seeds from the chickens and decompose into compost. I don't think

I have to tell you that the best to tool for doing the mowing is this one right here." He held up a kama, the familiar Japanese hand sickle. We laughed since we were all too familiar with that particular tool. No farmer in Japan goes into the fields without one.

"Many of the vegetables we sow today will not survive, but many will become established and thrive. Sometimes you need to cut the ground cover back once or twice to help the seedlings get off to a good start, but usually even that isn't necessary. The vegetables that do survive will flower, drop seeds, and reappear even more vigorously the next year. With each generation the plants revert more and more to the character of their wild ancestors, and many of them grow absurdly large, like those daikon over there." He pointed to a patch of daikon and asked Kin-chan to pull one up. It was, indeed, *huge*.

"You don't need to use compost. A mixed ground cover growing with microorganisms in a healthy soil makes compost all by itself.* What could be easier? You don't have to go to the trouble of collecting the materials, mixing them together over and over, and then spreading the compost on the field. Let the plants do the work. It helps if you have chickens and ducks grazing in the orchard as well. Plants, animals, and microorganisms living together is the best method for improving the soil. It is what nature intended.

"Enta-san, what do you think the overall design of our garden should be? Where should we plant the cucumbers and these leafy greens?" "I have no idea, Sensei," he replied. Sensei smiled and said, "I can tell that you have been living here for a while. Most people would say something like, 'Well, cucumbers need plenty of sunshine and well-drained soil so they should go here. Greens like the shade so we'll put them over there near those trees,' and so forth. We are going to take a completely different approach. We are going to mix the seeds of all the vegetables, ground cover plants, flowers, and perennials, and just toss them out together. The daikon will come up in the place that is good for daikon, and the mustard in a good place for mustard. The

---

* Among the most common ground cover plants in Mr. Fukuoka's orchard were white clover, alfalfa, vetch, lupine, buckwheat, rye, barley, mustard, radish, turnip, dandelion, daikon, burdock, comfrey, mugwort, and bracken fern.

seedlings will grow up in competition with other plants but they are also assisted by them.*

"If you use your head to decide where these plants should go you will be limiting nature's expression. If you start with the scientific trials it will take you a very long time to decide what to do and you will still make mistakes. But if you toss all the seeds together, all at once, in just one year you will know exactly where the best place is for the daikon, mustard, cucumber, burdock, or whatever. You don't have to worry about pH, drainage, sunlight, shade, or even the proper planting time. The seeds know all of that. Nature wants to work with us, if only we give it the chance by getting out of the way. It took me forty years to figure that out, but you can learn it all in just one or two seasons.

"In the first year everything comes up together and it looks quite messy. In the second year things become more settled; the vegetables naturalize, the weeds subside, and the clover becomes stronger. By the third year, the soil becomes revitalized, the plants start to move around, and unexpected things begin to happen. The daikon is here one year and the burdock is there. The seeds fall and the next year burdock comes up where the daikon was and the daikon moves somewhere else. Birds drop seeds and animals help to mix things up, too. It's quite wonderful to see. Each season brings new surprises. When you plan the garden yourself, you only get the result you expected. The fact that things do not move around in the typical garden is an indication that it is not a natural situation."

Sensei opened small bags and packages of seeds and emptied them into a large bowl. Then he walked over part of the area, scattering them

---

* This method of growing vegetables was developed by Mr. Fukuoka in accordance with his local conditions. Where he lives there is dependable rainfall throughout the growing season and a climate warm enough to grow vegetables all year long. Over the years he came to know which vegetables could grow among weeds and clover and the care they needed to thrive.

In most parts of North America and the world the specific method Mr. Fukuoka uses would be impractical. It is up to each farmer who would grow vegetables in a "semi-wild" manner to develop a technique appropriate to the land and the existing vegetation. There is a more complete description of Mr. Fukuoka's techniques for growing vegetables in *The Natural Way of Farming*.

into the weeds and clover. We each took a turn. "The reason I picked today to sow these seeds is because it looks like it is going to rain for the next few days. The soil is rich here so they shouldn't have any trouble germinating. If you don't know the right time to plant, or if the conditions are harsh, such as hard compacted soil, unpredictable rain, or having a lot of birds and rodents around, it is best to leave the seeds in their own pod or husk before sowing them. Sometimes it is helpful to coat the seeds in clay. I don't have to do that very much anymore, but at first I put vegetable seeds in clay pellets all the time. We'll talk more about clay pellets next week sometime.

"You don't have to do very much after that. Once you plant potatoes they'll come up year after year and outcompete the weeds. If you leave some in the ground when you harvest, you won't have to save seed potatoes over the winter. You don't have to stake the vines, just let them run along the ground, up the trees, or put out some bamboo or tree branches and let the vines crawl over them. It's the same with tomatoes and eggplants, just let them sprawl and grow into bushes. Grapes and kiwis don't need to be pruned; people just do that to get them to yield more. It doesn't help the plant at all.

"When you grow many different kinds of vegetables together on untilled ground you won't have any serious problems with diseases or insects. The plants are stronger if they are left to get along on their own. They taste better, are more nutritious, and can be used as medicine as well as food. It's easier, too. My wife keeps a traditional vegetable garden next to the house down below. It is convenient and fits well with village life, but it's a lot of work. Growing vegetables like wild plants takes a little more space but it is ideal for those who want to live a self-sufficient life. Any questions?" There were none. "Okay, finish cutting back the rest of the ground cover in this area and then go back to what you were doing before." We thanked him and he headed back down the mountain.

It was raining heavily the next day so we decided to stay in our huts until it cleared. It hadn't rained for a while so we were grateful for the freshness it brought and the feeling of renewal. Yu-san took the opportunity to bring out his magnificent collection of kimonos and work clothes. He told me about the different materials and the natural dyes that were used to make them, showing me the hidden pockets and compartments, and the elaborate ritual involved in putting them on and later folding them to put away. He topped it off by getting dressed in the formal winter samurai kimono, including a hakama, he had worn when he lived in the monastery. A hakama is a traditional Japanese skirt that is worn over a kimono. It is tied at the waist and drapes approximately to the ankle. He explained the various ties and sashes and the meaning of the seven pleats, two in the back and five in the front. Yu-san was quite an impressive sight sitting Japanese-style in front of the fire with the rain pouring down and thunder rumbling overhead.

"Remember we talked last night about how important things like duty and responsibility are to the Japanese people?" he asked. I told him that I did remember. "There is an even deeper awareness that goes right to the heart of the Japanese spirit. It is the concept of *mu*. Sometimes it is expressed as 'emptiness,' 'nothingness,' 'no-mind,' and in a lot of other ways, but the fact is, the more you talk about it, the farther you get from its true meaning. It is a perception that only occurs when the mind is completely tranquil. Then, without effort, your sense of being a distinct entity disappears and you become united with the universal consciousness. Thoughts keep coming, the body moves mechanically and gracefully, but the sensation is one of detachment and utter peace.

"Sensei sometimes refers to natural farming as *mu* agriculture. That mainly has to do with the state of mind of the farmer when working in the fields. He says that living as a farmer is no guarantee that you will ever have that experience, but by living close to nature each day, farmers have many opportunities for greater awareness. The key is to live humbly."

In the afternoon we gave the kitchen a complete cleaning for the first time in weeks. In the evening some shochu, a distilled beverage made from sweet potatoes, somehow made its way up to the mountain. Hide-san cooked up a batch of tempura and we sat up late singing Japanese and American folk songs.

Orchard Design and Seed Balls

A few days later Sensei came up to the mountain with more seeds. He called us together in a clearing overlooking one of the terraced hillsides. "Today I thought we'd talk about the structure of the orchard, and in the afternoon we'll put some of these seeds into clay pellets.* The location of this orchard is perfect for a natural farm," he began. "It is in the foothills where the weather is pleasant, and there is little danger from flooding. The surrounding woodlands have wild plants and mushrooms, and we get the advantage of the fertilizer that runs down the hillsides with the rain. When I started creating a natural farm here the soil was hard and depleted and there were only a few types of plants growing. It took me a while to change that, but now the orchard and woodlands provide everything we need to live a pleasant and comfortable life.

"I brought you to this particular spot because we have a broad view of the orchard as it is now. There are deciduous and evergreen trees of all sizes growing together with shrubs, berries, vegetables, and ground cover plants. There are also kiwis, grapes, and other vines growing up into the trees." He pointed to a large apple tree nearby that had ginger and wild onions growing near the base, a blueberry bush above them, kiwi growing on the branches of the tree, and an akebia vine growing above the kiwi. "See? Five different plants, all in the same place. This three-dimensional use of space is entirely different from the conventional orchard, which is designed with the commercial production of only one crop in mind.

"Besides having nitrogen-fixing plants like white clover growing close to the ground, it is also important to interplant nitrogen-fixing trees and shrubs like acacia, myrtle, locust, and alder to enrich the deeper layers of the soil. The acacia trees you see all around grow so quickly that in eight or nine years I cut them down and use them for firewood and for building. For each one I cut down, I plant others in different parts of the orchard. Trees that grow in a wild setting like this are much stronger than those that are grown in a conventional orchard, and they live longer, too. All I do is cut back the ground cover once or twice a year and keep tossing out seeds.

---

* Clay pellets, sometimes known as seed balls, or seed bombs, are small clay balls with seeds inside. Mr. Fukuoka did not invent the technique, farmers have been embedding seeds in clay pellets for years; he simply revived it. As a result, clay pellet seeding is again being used effectively on small-scale farms, for large rehabilitation projects, and in urban areas by guerrilla gardeners to seed particularly dry, compacted areas.

"I did not design this orchard. I *did* do a lot of terracing, especially on the steeper slopes to catch the runoff and make it easier to walk around. I also replaced some trees and I built a few simple huts, but mainly I just tossed out seeds and let the orchard take its own form. I mixed more than one hundred different kinds of seeds together and wrapped them in balls of clay. Then I scattered them everywhere. Each seed ball contained a miniature natural farm, its own universe, with trees, shrubs, vegetables, grains, flowers, and clover.

"Many years ago I sowed seed balls in one area that is now like a jungle. I suppose that most of the seeds did not survive, but the ones that did each made ten thousand more seeds. In a short time that area became the hillside paradise it is today. Over time, I scattered seed balls everywhere. When I started I had no idea what to do up here, so I asked nature to show me, and I offered seeds. Nature responded by creating this orchard."

We took a short break, then reconvened at a storage hut that had a flat open area in front of it. Sensei asked us to take out a sheet of plywood, some buckets, a few tarps, some bowls, and a frame with wire mesh attached to it. We already had a pile of clay we'd dug from a pit nearby. It was quite heavy and still a little wet from the recent rains. "Some of you have made seed balls before, but we could use more and this will give me a chance to show our more recent arrivals how to make them," he said, obviously referring to me, Kin-chan, and Tsune-san.

"First we'll take some of this clay and shovel it onto the plywood. Then we'll mix the seeds together in a bowl like this and scatter them onto the clay. Then we need someone to take off their shoes and pretend that they are crushing grapes in Italy. How about you, Enta-san?" Enta-san walked in the clay with his bare feet, the soil squishing up between his toes, while a few of us turned the clay from time to time. When the seeds were thoroughly mixed we pushed the mixture through the wire screen and onto the tarps below. The seed/clay mixture came out in small clods, which we rolled into balls with our hands. Each batch was set aside to dry.

"This method of using wet clay is simple and you can make quite a few seed balls in a short time, but if you need *a lot* of them, say, for seeding an entire mountainside, it is better to use a different method using powdered clay and a rotating drum, like a cement mixer with the partitions taken out. Then you mix the seeds together and put them into the drum as it is slowly turning. Add a little dried clay powder and spray a mist of water over them as the drum goes around. The clay will naturally adhere to the seeds. Then add a little more clay, a little more mist, more clay, and so forth. When the seed balls have grown to the proper size pour them out and let them dry until they are hard. It's amazing how many you can make that way in just one day."*

Tsune-san asked if you could use any kind of clay. "Red clay is best, the same as you would use for making bricks. You can find it everywhere around here, especially in the subsoil. Many people think of it as poor soil, but it actually has a lot of nutrients. The nutrients just need 'scissors' to cut them from the minerals. The scissors are acids that come with the rain and the exudates of plant roots and microorganisms."

"Should you add amendments to the clay?"

"I don't have to add anything here because the conditions are so benign. The soil is rich, there is rainfall throughout the year, and the seeds are concealed in the ground cover. In the rice fields the seeds are hidden by the succeeding crop that is already growing in the field when I sow the seeds. For seed balls in an average garden, I suppose you could add ashes from the fireplace or a little topsoil, but it isn't really necessary. We are mainly creating a shell that will protect the seeds until they can germinate and grow on their own."**

"How big should they be?" "How many seeds should be in each pellet?" "When should they be sown?" Questions kept coming in one after another until Sensei became a little annoyed. "Just put the seeds in clay, toss them out there, and step away. It doesn't really matter how big they are, how many or what type of seeds are in each one, or when you sow them. The important thing is that you actually do it. People

---

* There are a number of descriptions and videos of both methods on the Internet.

** The seed balls Mr. Fukuoka recommended for rehabilitating the human-caused deserts of the world using broad-scale aerial seeding are a bit more complicated to make since they have to be hardy enough to withstand the impact of being dropped from an airplane, extremely harsh conditions, and predation from birds, insects, and other animals. The method for making these super-strength seed balls is given in appendix D.

spend so much time trying to figure everything out first, and heaven help us if they decide to do a formal study, then nothing gets done. In the meantime the whole world is falling apart. It's better to let nature answer those questions for you, and much quicker."

I don't think Sensei was as much irritated as he was making a point, but either way our afternoon seminar ended rather abruptly. "Finish up making the rest of the seed balls and be sure to put them in the shed when you're done. It looks like it might rain again tonight." We thanked him, as we always did, and he walked off.

Kin-chan, the "golden boy," wanted to take a turn walking barefoot in the mud. After a few minutes he asked Hide-san if he was doing it correctly. Hide smiled and said, "There is no right or wrong on the earth, no fast or slow in the blue sky," an obvious takeoff on one of the sayings Sensei used from time to time. There was no disrespect meant—it was just funny to hear Hide saying it instead of Sensei, so we laughed and continued on until we had finished. Then we cleaned up, put the tools away, and headed back.

There was still time before dinner so I decided to walk to one of my favorite places, a clearing where you could look down on the valley, the fields, the village, and in the distance the Inland Sea. Everything was in miniature—the farmers working in the fields, the thatched-roof farmhouses, the occasional train, and the tiny cars streaming along the highway. A freighter glided out to sea from the port in Matsuyama. Everything seemed so busy. What was it that needed to be accomplished in such a hurry? It was a surreal sight, especially in view of what our lives were like in the orchard, and one that I have recalled often since I left the farm. It helps remind me of what is important in life and what only seems to be so.

It was starting to get dark so I turned and followed the path back through the trees. I could see the hut in the distance, illuminated by the flickering firelight, smell the aroma of dinner cooking, and hear the familiar sounds of friends talking and laughing together.

Science, Intuition, and Diet

Some days, usually on Sunday or when it was raining, Sensei would gather us together in one of the huts for a discussion about things like current events, the economics of farming, philosophy, or religion, but the topic he returned to most often was science. Because of his father's status in the village, Mr. Fukuoka had the opportunity to attend Gifu Agricultural College, where he studied modern techniques for research and large-scale farming. He specialized in plant pathology under the guidance of Dr. Makoto Hiura, one of the top scientists in the field. Dr. Hiura studied under the Harvard-educated and world-famous Dr. Kingo Miyabe, an ethnobotanist from Hokkaido. Dr. Miyabe designed the Koishikawa Botanic Garden, at Hokkaido University, in Sapporo. When it opened in 1886 it was only the second botanic garden in all Japan. The first, founded in 1684, is at Tokyo University. One of Dr. Miyabe's books, *Ainu Economic Plants*, which was published in 1923 and is still in print today, remains the standard text on the classification and uses of plants by the Ainu.*

With this background, and the experience of working for seven years as a research scientist, you might have thought that Mr. Fukuoka would have maintained a positive attitude about the usefulness of science, but he did not. For him, science revealed the fundamental problem in modern society—our alienation from nature. Since science is based on dividing nature into discrete entities, it can never understand the whole. When the results of scientific research are applied, especially in the natural sciences, mistakes are inevitable because the human intellect can never take all of the relevant interrelationships into account. The alternative, he suggested, is to experience the world as a unified, seamless reality, gaining knowledge directly through intuition and practical experience.

Most people believe that science, by applying an empirical methodology, is the only truly objective tool for understanding nature. Mr. Fukuoka, on the other hand, believed that science only leads humanity farther and farther from truth. It also creates an inflated sense of accomplishment, which leads to pride and overconfidence. "The irony," he said, "is that science has served only to show how small human knowledge is."

---

* An ethnic group of people indigenous to Hokkaido and the northern part of Honshu in northern Japan, as well as the Kurile Islands and the southern part of Sakhalin Island.

One day as we were working in the rice fields we noticed Sensei talking with a fellow wearing a starched white shirt and tie near the edge of the field. Scientists, reporters, and government officials frequently came to the farm so we thought nothing of it. About a week later the man returned, talked with Sensei for a while, and handed him a folder. Sensei had lunch with us that day and shared what the meeting was about. The man was a soil scientist from the local college. He had taken samples from Sensei's rice fields and analyzed them in his laboratory. The results, which were the contents of the folder, revealed a dangerously low level of phosphorous. He suggested that Sensei add more phosphorus to his fields as soon as possible.

"What do you think I should do?" he asked. By this time we had learned that when Sensei posed a question in that manner it was better to remain silent and wait for him to answer the question himself. He went on, "That fellow is basing his recommendation on laboratory results of the chemical composition of the soil, which he isolated from all other factors including the relationship between the soil and the plants. Then he concluded that we needed to fiddle with things in order to get the plants go grow better. He didn't even notice that the rice plants growing right in front of him were the healthiest rice plants in all Japan. The leaves even brushed his cheeks when he bent down to take the samples. I have a better way to analyze soil. I look at the plants. That usually tells me all I need to know. If I need to know more, I ask the plants and they tell me." He asked us again, "*Now* what do you think I should do?" More silence. Then he tossed the report into the fire.

Mr. Fukuoka did not use scientific experiments to set his course, but he did conduct trials to learn how to solve practical challenges. He had no attachment to the outcome and was not trying to understand how nature worked, he was simply asking for guidance. After observing the results he did not try to figure out how to fix the things that had not worked, rather, he saw what did succeed and went in that direction.

He also received guidance from another source, but it can hardly be considered a methodology. He talked directly with plants and animals and asked them what they needed and how they wanted to be treated. "We must look carefully at a rice plant and listen to what it tells us. Knowing what it says, we are able to observe the feelings of the rice as we grow it. However, 'to look at' or 'scrutinize' rice does not mean to view rice as the object, to observe or think about rice. One should essentially 'put oneself in the place of the rice.' In doing so, the self looking upon the rice plant vanishes. This is what it means to 'see and not examine and in not examining to know.'"[1] Once farmers truly enter into nature, *become* nature, this sort of understanding is possible. In that way, natural farmers have access to *far* more information than they would otherwise.

Another common topic of conversation was diet. Like everything else, Sensei's ideas about diet reflected his belief that nature should be the guide and human opinions should be ignored. The way we grew and prepared our food on the mountain demonstrated this philosophy and made it easy for us to understand. Almost all of the food we ate was grown on the farm or foraged nearby. We made our own miso, tofu, and pickles, and grew shiitake mushrooms from inoculated logs that were stacked in a shady area near the chicken coop. Our meals were prepared using simple culinary techniques to preserve the flavor of the food itself. Because the vegetables that grew in the orchard were grown like wild plants in clean, fertile soil they had a rich and complex flavor. The food was not only delicious, it *felt* good to eat.

A natural diet defines itself according to the local environment and the needs and physical constitution of each person. Animals know instinctively what constitutes a complete diet for them. When people became isolated from nature they lost that instinctive ability and came to rely on their intellect to figure out what they should eat. Today there are diets that are based on philosophy, religious beliefs, theories of human evolution, ethical considerations, and any number of other things.

Many people rely on Western nutritional science, which more or less promotes the idea that good health is maintained through a well-balanced diet consisting of a set amount of protein, starch, fat, minerals, and vitamins. It does not take into account the fact that each person is different, that they live in different climates, and that food is naturally available only during certain seasons. It is the same reasoning as that of the soil chemist who believes that healthy plants can be grown simply by providing chemical nutrients in the proper amounts regardless of the conditions they are grown in or the plant's relationship with the soil.

Following this scientific diet means that a full array of foods must always be available. To provide food that is out of season means it either needs to be grown in artificial conditions or brought from far away. Food grown in unnatural conditions, however, is often tasteless and contains fewer nutrients than food grown in healthy, well-balanced soil. Because the food we eat is not very nutritious we rely on supplements and vitamins to supply these nutrients. Our sense of taste has become less sensitive so strong flavors and sauces are used to keep the food from tasting bland.

During one discussion, Sensei explained that in a natural diet people select which foods to eat with their body, not their mind. "It is best to go with your natural feeling. If you feel like having comfrey, you just go out and find it. If you feel like having potatoes, or daikon, just walk into the garden and your hand will naturally find its way to the plant. In Japanese, the characters for 'feast' mean 'to run around.' That means to get food for a feast you need to go into the orchard and 'run around' to get potatoes from here, oranges from here, and burdock from over there. That is the real joy of the feast. As Christ said, all people have to do is go out and pick their foods as birds do in the wild."

Sensei concluded by saying that the main problem is that people have been separated from food production and diet has become separated from the human spirit. "When food, the body, the heart, and the mind become perfectly united within nature, a natural diet becomes possible . . . A natural person can achieve right diet because his instinct is in proper working order. He is satisfied with simple food; it is nutritious, tastes good, and is useful daily medicine. Food and the human spirit are united."[2]

The Rice Fields

The area near the hut where Yu-san and I lived was sometimes referred to as "downtown" because the kitchen was there, the chicken coop, the toolshed, and the cistern were nearby, and it was where to the road to the rice fields and the village began. One morning we were sitting in front of the hut processing a harvest of adzuki and soybeans when Sensei appeared with a crew from the local television station. They were going to do a story about Mr. Fukuoka, including a segment of him explaining the difference between natural farming and conventional agriculture to his students. They said it would only take twenty or thirty minutes.

Sensei brought an elaborate diagram with him, which he called the Natural Farming Mandala. It featured spiral arms flowing out from the center, making it look like the branching arms of a living organism.

The crew found a picturesque location, arranged a few battery-operated lights, and tested their sound equipment. When they were ready they pointed to Sensei and quietly said, "Go!" Sensei started by pointing to the center of the diagram, saying that it represented the source, the unmoving center. It was labeled with the character *mu*. He went on to say that the world is spinning out from the center in an ever-broadening spiral. "People have come to think that more, bigger, and faster are better, but this is only the idea of modern civilization. This way of thinking is motivated primarily by greed and the desire for material things. 'We must have progress.' they say. 'Without progress there is no meaning,' but is that really true? As we move out from the center, which is nature, as we approach the limits of growth, everything begins to separate and come apart, and the unity and tranquility of the unmoving center is lost. At the same time, some people react to this and feel the need to return to the source. They want to live and enjoy a simpler life—"

"Cut!"

First the microphone wasn't working properly. Then it was the camera. When the director thought everything was ready, he took Sensei aside and asked him to remember that the film was for a general audience, so please make the message a bit more understandable to the average person. Sensei promised that he would. They tried the shot again. Sensei picked up exactly where he left off, but again there were technical problems. Soon the situation devolved into a slapstick affair with one thing going wrong after another so they had to film the interview over and over again. We enjoyed the diversion, but after several hours Sensei let us go back to work while he continued on with them.

We met at the end of the day and told him how bad we felt leaving him alone to deal with the media people. Sensei smiled and thanked us for our concern, then said, "I've lived here for more than thirty years and no one ever came up here or seemed to care. Now all of a sudden everyone wants to see what I am doing. Perhaps we have reached the end of the period of expansion and have entered a new era of contraction. I hope so, anyway. At least you had a chance to work and enjoy yourselves for part of the afternoon anyway." Eventually, the program aired and had a good response.

On another occasion I asked Sensei about his relationship with the other farmers in the area and why no one else in his village had tried natural farming. He said that his interactions with his neighbors were cordial, but when they happened to meet each other on the way to the fields their conversation was usually superficial, like talking about the weather. "When farmers pass my fields, they see them but pretend they are not there. They have no idea what I am doing and are not interested in asking me about it. I suppose some have seen pictures of my orchard in the newspaper and on television, but no one has ever asked me to show them around." Mr. Fukuoka did not go out of his way to try to explain what he was doing to them, either. He never gave a public talk in the village or in the nearby town of Iyo.

He said that he and his neighbors were on divergent paths. There is nothing typical farmers treasure more than stability, so they use methods that are known to be reliable. If farmers have a bad year the effects are felt for a long time. If it continues for two or three years, they may not be able to make payments on their loans. In the old days a bad harvest or two meant people starved. Mr. Fukuoka took a different approach by trying all sorts of unconventional cropping methods. He never did the same thing twice. His fields were full of weeds, clover, insects, and straw, which was incomprehensible to the other farmers who were used to ordered rows of plants growing in a sterile, controlled environment. Also, Mr. Fukuoka had noticeably uneven yields while he was developing his method, including a few years that most farmers would have considered complete failures.

"Farmers around here play it safe by following the instructions of the government agricultural cooperative. They tell them what varieties to plant and when, what fertilizers to use and in what quantities, what type of pesticides to use and when to spray them. They won't do anything unless it has been thoroughly tested at an experiment station and then recommended by the cooperative. I have no doubt that if my method was tested and approved, farmers across the country would try it, but the few tests that have been done have always added scientific modifications and the researchers quickly lost interest. This is not surprising because natural farming, which uses no machines and no chemicals, is in direct conflict with the Agriculture Ministry's policy of encouraging industrial-style agriculture. Without official approval, farmers won't believe that it is possible to grow crops the way I do."

During Mr. Fukuoka's lifetime, natural farming was seen not so much as a curiosity by the agricultural establishment as it was a threat. It was never seriously investigated, partly because if natural farming was shown to have merit by scientific standards it would demonstrate not only that conventional agriculture was not, in fact, the only correct way to farm, but also that it was not even the *best* way. That would challenge their assumptions to the core. Mr. Fukuoka was vilified, not for being wrong, but for being different, and being different is not something the agricultural orthodoxy of Japan could tolerate.

A few days later we were working in the rice fields. Sensei came by and said he had a few things to show us. As we walked through the fields, green spotted frogs, moths, crickets, dragonflies, and spiders came to life in a flurry. Lizards, mice, and snakes raced over the surface of the soil, and moles, gophers, and earthworms burrowed below. We followed Sensei to a relatively large section of the field where a leafhopper called unka, a tiny cicada, had ruined most of the crop by breeding at the base of the plants and sucking the plants' sap. There were smaller patches with similar damage here and there in other parts of the field.

"This insect is considered the most difficult pest in rice-growing," he said. "Everyone else sprays four or five times during the growing season to control it, but I don't mind having it in my fields. It thins the weakest plants and allows more sunlight to reach the others. I don't think it reduces the yield very much, it may even increase it for all I know, but it never gets out of hand. There are so many other insects living nearby that when the population of unka grows, so does the population of its predators. Nature provides all the insect control I need."

As we continued walking, Sensei proudly pointed out some of the plant diseases that also made their home among the rice plants, mostly various forms of fungi. "This one"—he pointed to a bright-orange fungus growing on one of the leaves—"is called *ine koji byo* [rice starter disease]. You can take it right out of the field and use it as a starter for making soy sauce or miso. The plants are strong enough to tolerate it most of the time."

He showed us how to tell the history of the growing season by looking at the nodes of a single plant, then went on to explain how new varieties of rice appear in his fields without his doing anything. "Grasshoppers and cicadas make small holes in the grains of rice just as the heads are developing, then snails, slugs, and cutworms eat them down to the stamens as they crawl over them at night. Windblown pollen adheres to the flowers, creating cross-fertilization. Rice is normally thought of as being self-pollinating, but it can also be pollinated by other plants, which creates new varieties. I used to breed new varieties myself, but now I don't bother. This natural cross-fertilization cannot happen in conventional fields because they are spraying all the time."

Next we walked to the edge of one of those conventional fields. He pulled up one of the plants and then dug one up from his field to show us the difference in the root systems. The roots from his neighbor's field were small, shallow, kind of slimy, and had a dark color; those from Sensei's field were very long, thick, and had large white branches. The volume of roots from Sensei's plants looked to be about four or five times greater.

"Rice can be grown in a flooded field, but it doesn't have to be; in fact, it grows better without flooding. People started flooding the fields hundreds of years ago mainly to control weeds. By now they have forgotten why they do it, thinking that they have to in order to get a crop. One look at the roots of these plants should be all you need to know about which method the rice plant prefers. The roots are larger and more robust because they have to grow deep to find water. Also, it is fed by the microorganisms in the soil and has access to more oxygen and micronutrients. Plants, soil, and microorganisms form a sacred alliance."

Mr. Fukuoka got the idea for his rice growing when he happened to pass a rice field that had recently been harvested. There he saw rice seedlings growing up voluntarily among the weeds and straw. From that time on he stopped plowing, stopped flooding the paddies, and stopped growing seedlings in a started bed in the spring and then transplanting the young shoots to the main field. Instead he broadcast the seeds directly onto the surface of the field in the autumn when they would naturally have fallen to the ground. He learned to control weeds by spreading straw and growing a permanent ground cover of white clover. As we looked out at the ripening field of rice, Sensei said, "Most farmers believe that they are responsible for growing crops, but I didn't grow this crop of rice. Only nature can create and grow crops. All I did was sow the seeds and spread straw."

Red spider lilies* had just begun blooming on the paths between the rice fields. When these flowers appear, no matter how hot it is, it is a sign that autumn is on the way. Soon the rice harvest would begin. We stopped to sample some figs from a tree growing beside the path on the way back to the mountain. The evenings were chilly now, and leaves and twigs had started to accumulate on the surface of the pond.

There was only one full day's work left in the rice fields before the harvest and the orchard was in pretty good shape, so Sensei gave us a rare day off. It was such a remarkable event that after breakfast we sat there staring at one another not knowing what to do. Gaki-san, a relatively recent arrival, and I decided to walk to the coast about five miles away.

We walked for a while along the hot, dusty highway crowded with families out for a Sunday excursion, then walked on paths through the rice fields. We avoided the fields where farmers were spraying their crop. When we finally did make it to the coast, we could see that the water was too polluted to swim in. There were piles of old fishing nets, sake bottles, and trash piled along the shore, and rotting fishing boats that no longer went to sea because the waters had been fished out long ago. It was depressing.

We decided to stop at a noodle shop for lunch. Gaki-san brought up something Sensei had talked about a few days earlier: "People say that we are living in a fantasy world up here on the mountain, but it seems to me they are the ones who are living in a dream. Go into a typical noodle shop sometime and look around. There is music coming out of a jukebox or a television set, menus are hanging on the wall, a canary is in a cage in the corner, and people are sitting under neon lights in starched shirts and ties slurping noodles. It's the most everyday scene there is, right? Well, I think that what most people consider 'everyday' and 'normal' is the weirdest reality of all." We laughed when we looked around realized that we were sitting in that very noodle shop!

* *Higanbana*, or "equinox flower" in Japanese.

With our spirits lifted somewhat we headed back to the farm. Gaki-san said, "You know, Larry-san, this wasn't really such a bad day. Remember how adorable that young girl was when she smiled and waved to us as she pedaled by on her bicycle? And the snake we saw swimming in the irrigation canal with a dragonfly on its head? Just those things alone were enough to make this day enjoyable for me." It was dark and we were tired by the time we got back. It was *so* good to be home.

The One-Straw Revolution

The rice harvest was on. We cut the grasses with a special serrated sickle and left them on the field for several days to dry. The neighbors hung their rice on wooden racks to get it off the muddy ground, but Sensei's fields were already covered with winter barley so we just spread the rice over it. When the crop was dry we threshed the grain using a foot-powered rotary drum with wire pegs on it. Then we sowed the next year's rice crop along with some clover seeds, and spread the uncut rice straw over the field. The young barley plants were trampled in the process but quickly recovered.

One afternoon while we were threshing rice in the courtyard of his home in the village, Sensei came out of the house with a big smile on his face. He was holding the copy of the Japanese edition of *The One-Straw Revolution* he had just received from his publisher. After reading the book, Tsune-san and I decided to translate it into English and try to get it published in the United States. Neither of us had experience writing, editing, or doing translation, but we felt that it was important to make Sensei's point of view and his example of a viable organic no-till agricultural system available to people outside Japan.

This was before word processors or personal computers, so the first thing we needed to do was get the old typewriter we found in one of the huts into working condition. It had no ribbon, several keys were missing, and the carriage had an irritating habit of sticking on the return. I took the train to Matsuyama city several times to get it repaired, usually stopping to visit Matsuyama Castle and the nearby public hot springs along the way.

We enlisted the help of Chris Pearce, a friend I met when we lived together at the ashram on Suwanose. Chris grew up in Japan and could read and speak Japanese fluently. He gave us a literal, first-draft translation. Chris was not a farmer and had never been to Mr. Fukuoka's farm so some sections of the manuscript were difficult to understand. We were grateful to have the book in English but there was still a lot of work to do.

By that time we were in the middle of the four-month citrus harvest, my favorite time of year. We climbed into the trees or used ladders, picked the fruit, and put it into buckets we carried up with us. When the buckets were full we poured them into crates, then loaded the crates into the bed of a small pickup truck that shuttled between the orchard and the sorting shed in the village. Most of the fruit was considered acceptable for shipment even if it had a few blemishes. The ones that were seriously bruised were put into a separate area for "seconds" and were sent to the local cooperative facility, where they were made into juice.

One of the things I liked about harvesting the mandarin oranges was being up in the trees. It was fun, and you couldn't help but notice the beauty of the green leaves, the orange fruit, and the blue sky. Another was the camaraderie. Since it was such a big job, Sensei's wife, Ayako-san; his son, Masato-san; Masato's wife, Rieko-san; and several others from the village also pitched in. It made the work easier, and we all enjoyed the infusion of new personalities, new stories, and new jokes. No one ever felt rushed; the work proceeded at an even pace with regular breaks for tea and snacks. It was gratifying to look back at the end of the day and see how much we had accomplished.

While we were working, Tsune-san and I carried on a nearly continuous conversation about the manuscript and the questions we had about it. During breaks I noted these questions in a small spiral notebook I carried with me. Then, after working all day, we met with Sensei several times a week to clarify those passages. Sometimes the meetings took place in Sensei's living room in the village, but more often they were held in the sorting shed, which was across the road. "What exactly was the point of this story? What *are* the Seven Herbs of Spring? How do you know when to sow the vegetable seeds in the spring and fall? Did you really mean this literally, or were you just trying to help people see things in a different way?" Sometimes our discussions went far into the night.

We also talked about the differences between the Japanese and English languages and the differences between Eastern and Western readers. For example, the Japanese way of telling a story or making a complex argument is different from the approach that is generally taken in English. In Japanese, the author typically begins with the theme of the story or the point of the argument, then offers anecdotes or arguments that support the theme, which is restated each time. This circular style can soon become tedious for Western readers who are used to a more linear approach that has a beginning and moves directly, step by step, to the conclusion. We agreed that it was important to restructure some sections of the book so they would be as clear and natural for an English-speaking audience as the original is for Japanese. Footnotes would also be needed to explain terms and expressions that required a cultural context English-language readers might not be familiar with, as well as an introduction. When we finally had a manuscript we were all satisfied with, I was entrusted to take it to the United States to find a publisher.

I was a little sad as I walked down the winding road from the mountain for the last time, but I was also exhilarated. The two years I spent living on the farm had transformed me. At that moment I felt like I was the luckiest person in the world. Mr. Fukuoka met me in the courtyard of his home. We laughed and talked, and he wished me well. Then I swung my backpack onto my shoulders and headed for the train station.

CHAPTER 4

Traveling with Mr. Fukuoka in the United States

I HAD NO IDEA WHERE TO BEGIN. I knew *nothing* about publishing or what was involved in finding a publisher. Also, I knew that the manuscript, even with all our work, was still rather crude and amateurish.

I decided to show it to my friend John who at the time was the managing editor for the University of California Press. His response was a real eye-opener. He said, "Wow, Larry. This is terrific! I don't think you have any idea what you've got here. It needs some work, but I'm sure I can get UC Press to publish it. I think you may be able to do better, though." This is when I first realized that the book would find the right publisher all by itself. I decided to toss out a few seeds and see what happened, just as Sensei did at his farm.

The next person I showed the manuscript to, was Gary Snyder, a poet and environmental activist who lived in the foothills of the Sierra Nevada. He had recently returned from several years in Japan, where he studied Zen Buddhism and was instrumental in founding the ashram on Suwanose Island. I approached him after a poetry reading he gave in Berkeley one evening and asked if he would have a look at it. He had heard of Fukuoka-sensei and was glad that one of his books had been translated, but said that he was not the appropriate person to evaluate it. He said he was more of a mountain person than a farmer, but "why don't you send it to my friend Wendell. Here's his address in Kentucky. Tell him I suggested that you send him the manuscript."

I have to admit that at the time, I didn't know who Wendell Berry was, but since Gary suggested that I send the book to him, I did. Wendell called on the telephone about a week later. After chatting for a few minutes he asked me if I had found a publisher yet. I told him that I had not. He said, "If you will allow me, I think I can help you with that." I gladly accepted his offer, and from then on this kindly southern gentleman took the book under his wing and made sure that everything went right with it.

Despite the unprofessional quality of the draft, Wendell was impressed with the content, especially because Mr. Fukuoka was a farmer who had verified his ideas in his own fields before offering his advice to other people. He suggested that we approach Rodale Press, partly because he did not want the book to be pigeonholed as a "new age" work. He wanted to be sure that it would end up in the hands of real farmers. Rodale Press published *Organic Gardening* magazine, which had a very large circulation as well as a popular book club that reached hundreds of thousands of farmers in the heartland of America. Along with *The Mother Earth News*, they were the leading source of inspiration and practical information for the growing environmental and back-to-the-land movements.

The contract was signed and Wendell became Rodale's editor. Over the next year or so we worked together by mail to make the book as clear and as practical for American readers as we could. Besides cleaning up the language, we had many issues to discuss and decide. Would we use metric or English measures? "Mr. Fukuoka seems to contradict himself when he says this on page fourteen, but that on page thirty-one." "When farmers in most parts of the United States read how he grows vegetables they'll think it is impossible because their climate is so different from his. We'll need a footnote to explain that." When Wendell made stylistic changes he always explained *why* he made them. By taking the time to do that, he taught me how to write effectively. He also pointed out the connections among writing, editing, and clarity of thought, and for that I will be forever grateful.

It was unusual for Rodale Press to publish a book like *The One-Straw Revolution* because it was at once practical, philosophical, and spiritual. It was also written with an elegant literary style. To reflect this, Wendell and I believed that it needed a distinctive design. For all of its strengths, Rodale Press did not excel in book design. This was understandable since most of their publications were straightforward, giving practical how-to-do-it advice and information. Wendell lobbied for and received permission for us to do the book design ourselves, rather than Rodale handling it in-house, as was typically done.

Jack Shoemaker, Wendell's longtime friend and literary agent, had a bookstore and small publishing company called Sand Dollar Books on Solano Avenue just north of Berkeley. They mainly published finely crafted, hand-typeset, limited-edition books of poetry and short essays. There were a number of small poetry presses like that in the Bay Area at the time. Jack put together a team from those quality small presses with George Mattingly as the lead designer. This was a wonderful development for me because I got to see all that is required to produce a book from beginning to end.

I met with the illustrator Chuck Miller at his home in nearby Albany. We knew that each chapter needed a full-page drawing so I loaned him some photos of the Japanese countryside and Sensei's farm to provide inspiration. He also created a number of small designs and whimsical motifs that were sprinkled here and there throughout the book. We needed to use special paper that was not too porous so it would hold the fine lines of his intricate pen-and-ink drawings. Just as I was leaving our initial visit I asked Chuck if he could do a drawing of a heron. He said he would be delighted and told me that he began his career as a wildlife artist. The heron he drew appears on page nine of *The One-Straw Revolution.*

Howard Harrison traveled to Japan to take black-and-white photos of Sensei and his farm to supplement the ones we already had. Later we went over the proofs together and selected the ones we would include. I chatted with David Bullen as he laid out the pages in his home studio in Berkeley and visited typesetter Bob Sibley in San Francisco. That was the first time I got a look at that miraculous new invention called "word processing." It reminded me of the countless hours it had taken to retype the manuscript each time the handwritten changes became too numerous to go on without a clean draft. It took a little more than three days with three of us typing in two- or three-hour shifts.

Finally, after almost two years, the book went off to the printer in Chicago. The wait was excruciating, but then one day, there it was, sitting in the mailbox at my parents' home in Los Angeles. Pretty soon it started showing up in bookstores and the reviews came rolling in. The book was going to be a hit.

Shortly after the book was published I visited Wendell at his farm in Kentucky. He and his wife, Tanya, were gracious enough to let me stay with them for a while, then I moved to a neighbor's farm and helped him bale hay, plant tobacco, take care of the cattle, and work in the vegetable garden. One day there were thunderstorms and the grass was too wet to bale so we drove to Louisville and spent the day watching horse races at Churchill Downs.

While I was there, Wendell told me about the problems facing small-scale diversified farmers in his community and throughout the United States, things I was not aware of since most of my farming experience had been in Japan. What he described, however, was similar to the challenges faced by local farmers and farming communities in Japan and around the world. We also talked about Sensei's philosophy and farming techniques and the similarities and differences with his experience in Kentucky. Besides farming using his team of horses, Wendell was teaching English at the University of Kentucky at the time. Just before I left he said, "You know Larry, you are an English teacher's dream. Most of my students want to become writers but they're all about style and don't have much to write about. You had something important to say, you just didn't know how to say it."

"Big Trees" and the Camp at French Meadows

The publication of *The One-Straw Revolution* in the United States marked a turning point in Mr. Fukuoka's life. For more than thirty years he had labored in his small village in relative obscurity. After the book's publication he was suddenly known and respected throughout the world, and he started receiving invitations to visit from supporters everywhere.

The first invitation arrived in 1979 from Herman and Cornelia Aihara, macrobiotic teachers in California who sponsored a summer camp near the reservoir at French Meadows in the Sierra Nevada Mountains. Mr. Fukuoka accepted and came to the United States with his wife, Ayako-san, for a six-week visit that included stops at farms, farmers markets, universities, and natural areas in California, the Northeast, and New England.

The Fukuokas arrived in San Francisco with surprisingly little in the way of luggage. We made our way across the bay to Berkeley and the home of my friends Mino and Fusako, where we stayed for a few days to allow them to recuperate from the long flight. During that time Mr. Fukuoka continuously talked about how remarkable it was to see the earth from the air. He was surprised when he looked down on the California landscape for the first time. It looked like a desert to him and not at all like what he was expecting. He couldn't wait to investigate further.

We headed off, bright and early, in my small Honda Civic. I was excited but also a bit nervous since I had planned the itinerary, made the arrangements, and taken on responsibility for making sure the Fukuokas were comfortable. I needn't have worried, though. Sensei and Ayako-san were terrific travelers, game for anything, never complained, and spoke up right away whenever they needed something. They were extremely friendly and outgoing toward everyone they met. As it turned out, my biggest challenge was keeping up with them. Sensei required very little sleep. By the time I woke each morning, he had already been up for some time, had made himself a pot of tea, and was eager to be briefed on the day's activities.

We passed through the arid brown hills separating the Bay Area from the Central Valley. When we came over the pass, the broad Sacramento Valley came into view with the snowy peaks of the Sierra Nevada far in the distance. This seemed to take Sensei's breath away. He had never seen an agricultural valley anywhere near that large. In fact, there were many things that seemed really, really big to him—the size of the trucks on the highway, the farm machinery, the houses, the people, and the ample portions of food served in restaurants. It reminded me of how small everything had seemed to me when I first arrived in Japan.

We continued across the valley, past almond, peach, and apricot orchards growing on the valley terraces and fields of tomatoes, sunflowers, and alfalfa in the bottomlands. Then we climbed into the foothills of the Sierra Nevada on the east side of the valley, and finally to higher elevations with impressive forests of ponderosa, lodgepole, and sugar pines, incense cedar, California black oak, and fir trees. Shortly before we reached the campground at French Meadows, we stopped at a botanic area maintained by the USDA Forest Service where the northernmost grove of giant sequoias* was growing. There was a small parking lot and a self-guided loop trail of about half a mile.

There were only six "big trees" in the small grove, but they were large and impressive. For some reason the area was not covered by ice during the last glaciation, so the trees and associated plants have remained largely the same as the vegetation that existed there tens of thousands of years ago. The understory plants consisted of dogwood, currants, manzanita, ceanothus, California nutmeg, and azaleas. It was a lovely spot.

We got out of the car and made our way to the trail. We came to the first redwood trees almost immediately. Sensei looked up at them and then he briefly glanced at the understory. From then on he only paid attention to the forest floor. He asked to use the hand magnifying glass I brought along and examined the lichen and mosses that were growing on the granite boulders. Then he got down on his hands and knees and started digging in the duff. He smelled it, studied it, and then pointed out the different kinds of fungi he found living there.

After a few more minutes he said, "This forest is the same as the virgin forests of Japan. The plant species are closely related, although different, but the mosses and lichens, and the life in the litter layer, are identical."

I asked, "Wouldn't all old-growth forests be similar?"

"In many ways yes, but in this case the species of ancient life-forms are *exactly* the same." He added, "Besides, it *feels* the same," as if that should be proof enough to settle the matter, and for me, it was. I understand that others may have needed more evidence to have accepted Sensei's conclusion, but I had seen how comfortable he was with nature at his own farm, and had complete confidence in his ability to be at ease with nature anywhere. He had only been in the United States for a few days and had spent perhaps twenty minutes at that particular location, but he seemed right at home, instinctively understanding the spirit of that particular forest.

---

* *Sequoiadendron giganteum.*

I was raised in California and did all my soils fieldwork there. I thought I had a fair understanding of the landscape, but sometimes you can get a little too comfortable with conventional interpretations and never think to contest them. Having Sensei come and see the California landscape for the first time was a reality check for me, and it turned out that his opinions consistently challenged what many Californians considered to be indisputable truth. He saw things that we could not because we were too close to them. This brief visit to the Big Trees Grove in the Tahoe National Forest gave me a glimpse of how interesting traveling with him would be. I learned to give him few details before we came to a new place, just listen and learn from his first impressions.

Herman and Cornelia Aihara greeted us as we drove into the French Meadows Camp. When word got around that Sensei had arrived all of the other campers came over as well. He was surprised and visibly moved by this outpouring of adoration. All those years he had lived and toiled in relative anonymity in Japan, always being criticized, always put on the defensive. But here in the United States he was a star surrounded by almost one hundred well-wishers who were receptive and wanted to learn from his experience.

Everyone was camping. A stream flowed through the campsite to the nearby reservoir, and the air was filled with the fragrance of the evergreen trees. There was a small stage near a campfire circle, a makeshift outdoor kitchen, clothes hanging to dry on tree branches, and bright smiles everywhere. Sensei spoke to the entire camp in the mornings and chatted informally with small groups of students in the afternoon. Sometimes he would walk by himself along the creek. We stayed for nearly a week so there was plenty of time to unwind and take it all in.

Herman and Cornelia Aihara were two of the warmest people one could ever hope to meet. Their mission was simply to help others live healthy and happy lives. They were students of George Ohsawa, "the father of modern macrobiotics," and came to the United States in the early 1970s to introduce his teachings here. They opened the Vega Study Center in their home in the small town of Oroville, California, where they gave classes, offered workshops, and published books and a newsletter. They were practical, down-to-earth people who were always accessible. Like Sensei, Ohsawa was virtually ignored in Japan, but when his message spread to the United States through the teaching of the Aiharas, and Michio and Aveline Kushi in Boston, it quickly became well known worldwide.

Sensei's talks at French Meadows centered on diet. He noted that Ohsawa's "Path to Nutrition," known in the West as macrobiotics, is based on the concepts of yin and yang, but in its application is almost identical to the diet of traditional rural Japanese people. He said that Western natural science, including nutritional science, grew out of discriminating knowledge, as did the Eastern philosophy of yin and yang and the I Ching. For that reason, neither can embody absolute truth because both are only an interpretation of the world. To achieve a true natural diet you must abandon the world of relativity. He asked the students not to become attached to the details of the teaching itself because that would lead to confusion—eating with the head, not with the body.

He said that in a world where people have become estranged from nature they lose their ability to know what to eat to achieve a healthy body and spirit. Under those circumstances, yin and yang may serve as a temporary aid, a guidepost toward restoring order, but that should not be considered the ultimate goal. "When the individual is able to enter a world in which the two aspects of yin and yang return to their original unity, the mission of these symbols comes to an end."[1]

At the time I, too, had the impression that macrobiotics was a complicated system of balancing the forces of yin and yang, expansion and contraction, and acid and alkaline. After sitting in on a few talks by Herman Aihara, however, I learned that that is only true in the early stages of the practice. Mr. Aihara said that macrobiotics is not only about establishing physical health, but also about leading a life that is in tune with the natural order. Then your self-image improves, relationships with other people become richer, and you can enjoy a more positive attitude toward life in general.

He said that macrobiotics is mainly about attaining freedom in spirit. That comes about by recognizing and accepting physical limitations. We are free to think anything we want, so we think we can eat anything we want and still maintain good health, but we must learn to discipline ourselves to accept and live within the physical limitations God has given us. When we think we can do anything we want, we become arrogant, and that leads to sickness.

When someone is able to refine their judgment to the point that they can instinctively choose what their body needs, they no longer have to rely on the principles. In other words, your spirit has become realigned and you can trust yourself to make good decisions without having to think about it. Then the mission of macrobiotics has been accomplished. None of this is possible, however, unless you maintain a sense of humility.

A farewell campfire was held in Sensei's honor the night before we left with lots of singing and good cheer. It was hard to leave this idyllic gathering, but in the morning, with a deep bow, he said good-bye to everyone and we continued on.

The Olala Farm

Sensei's next scheduled event was a farm near the small town of North San Juan, in Gold Country.* We were not expected until late afternoon so we decided to make a few stops along the way. The first was at an organic truck farm where several families lived together growing organic vegetables that they sold at the farmers market in Davis and to restaurants in Sacramento. The conversation mainly centered on how difficult it was for small-scale organic farmers to make a living. The second stop was at a conventional almond orchard. The grower took us to see the trees that were laid out in a perfect grid. The surface of the soil was completely clear of weeds, giving it the appearance of a pool table. The farmer explained that the almonds were harvested using a machine that shook the trees, causing the almonds to fall to the ground. Keeping the surface clear of obstructions facilitated that process.

What interested me about these visits was the way Sensei interacted with these farmers. He was friendly and outgoing, and did not bring up his own experiences or offer advice unless specifically asked to do so. He seemed much more interested in the people themselves than their farming techniques. I had always thought of Sensei as a shy person who was a bit awkward socially, but that turned out not to be the case at all.

---

* The region along the foothills of the Sierra Nevada where the California Gold Rush occurred, beginning in 1849.

During the course of the tour I came to appreciate his kindness and his humanity. It did not matter if you were a farmer, an artist, a plumber, or a sales associate at the local Sears store, he wanted to know about *you*. When he gave a talk in front of a large group he was a fiery crusader for the protection of nature; in a smaller group he had the same message, but was more relaxed and personal. One on one, he was everyman.

Sensei was also extremely tolerant. I saw that the first time I met him, when he did not flinch at my beard and long hair, and later when he did not send people away from his farm simply because they had no farming experience. Once, when we were sitting together in a café in Berkeley, a fellow with a spiky blue Mohawk walked in wearing black leather and chains. I don't remember exactly what I said at the time, but it must have been something like, "Well, it takes all kinds." Sensei was furious. "Never judge someone simply by their appearance," he said. "You have no idea who a person is until you get to know them." He went on to tell several stories about people he had met in the past that illustrated his point, shifted his chair so we no longer faced each other, and did not talk to me for the next half hour.

We headed north from Nevada City on Highway 49, crossed the Yuba River, and continued on to San Juan Ridge, the site of a Gold Rush hydraulic mining operation. By 1851 most of the gold in the nearby streams had already been found, but miners discovered that there was also gold in ancient riverbeds that had been covered by soil, sometimes several hundred feet deep. Water from elaborate wooden flumes ran into hoses outfitted with nozzles to create cannon-like streams of water so strong they could kill a person from two hundred feet away. This force was used to wash away the soil, exposing the gravel lode beds below. The gravel was then cleaned in sluices to separate it from the gold. It was an early strip-mining system in which all the sediment and debris was deposited into the Yuba River. At the nearby Malakoff Diggins,* forty-one million yards of material was removed, creating a canyon seven thousand feet long, thirty-four hundred feet wide, and nearly six hundred feet deep.

---

* The term *diggins*, or *diggings*, refers to the tailings that were left behind at the end of the operation.

The sediment clogged the river channels, raising their beds. Marysville, Yuba City, and Sacramento flooded repeatedly as a result. The silt washed all the way to San Francisco Bay. One landowner, Edward Woodruff of Marysville, finally filed suit against the North Bloomfield Mining and Gravel Company in 1882. In one of the first environmental protection decisions ever issued, the company was ordered to stop using the hydraulic mining method. In 1893 the US Congress passed the Caminetti Act, which authorized the US Army Corps of Engineers to only issue licenses to individual mines that could demonstrate that their debris would not be discharged into waterways, effectively ending the practice.

We sat on a log and looked out at the scene. Even after more than one hundred years the area resembled a barren moonscape with piles of white rocks and gravel. The only living things we saw were a few stunted pine trees and a lizard sunning himself on a rock. The glare was unbearable. "This," Sensei said, "is a magnificent monument to human greed."

We drove up the dusty dirt road leading to the Olala farmhouse where our hosts, Robyn and Arlo, were on hand to greet us. They had been active in the San Francisco art community until they decided to move to the country to become farmers in the early 1970s. A number of writers, poets, artists, musicians, and assorted free spirits all moved to the area at about the same time, making it one of the more eclectic rural communities in California. Many of them built off-the-grid houses and planted vegetable gardens. At the time Robyn and Arlo were the only ones who had taken up farming.

About forty or fifty people gathered in a meadow in the late afternoon to hear Sensei speak. At one point he described how he had improved the depleted soil in his orchard by growing a continuous ground cover of plants like burdock, radish, dandelion, white clover, and acacia trees. He liked these species because they were vigorous and produced many strong roots that aerated the soil and recirculated nutrients from deep in the ground. He acknowledged that in one situation or another all of these plants were considered too aggressive to manage easily. The way he dealt with that was to scatter the seeds of all of them together and let the plants work it out among themselves. Some did well in one area while others thrived in other places. Sometimes they all grew together.

The pattern changed from year to year. He referred to these plants as "hard workers."

He went on to say that while driving around in the Sierra Nevada foothills he had seen a lot of damaged land, but also noticed that several plants that grew almost everywhere were effectively helping the soil recover. Then he turned and pointed to plants growing on the hillside behind him. "This one is my favorite." There was an audible groan from the audience* and then some giddy laughter. "What did I say?" he asked. The hillside Sensei had pointed to was covered with Scotch broom,** the plant the locals generally considered their most noxious weed. What followed was a spirited and sometimes heated discussion about native versus non-native species, "invasive" plants, and human responsibility.

One fellow, who was a native plant enthusiast, pointed out that Scotch broom was an invasive, non-native species that had "taken over," pushing out other plants that were native to the area. He thought it was a menace that should be eradicated. Mr. Fukuoka replied by asking him, "What is native?" He pointed out that species move around the world all the time even without people's help. "Seeds fly from place to place in the wind even to distant continents. Birds and other animals can also carry seeds long distances. The composition of plants in a given area was different ten thousand years ago than it is today, or one hundred years ago, or even yesterday. The coast redwood trees that people consider native to California have been here less than twenty million years. Before that, California had a subtropical rainforest. Those plants are now in Mexico and Central America. Everything is constantly in flux.

"Lately people have accelerated the process. We have consciously carried species from one place to another, and now, with people driving all the time, and with international travel and commerce, we are doing it unconsciously as well. When you walk through your garden and then get on a plane to Hawaii you are carrying who knows how many species of microorganisms on your shoes and seeds on your clothing? It

---

* On his second visit to the United States, Mr. Fukuoka heard similar groans from vegetable growers in the Puget Sound area when he suggested that they keep their fields covered with a continuous cover of white clover.

** *Cytisus scoparius.*

has already happened, and the spread of species around the world will only continue. Besides, even if we were to go to the trouble of replacing species like Scotch broom with native plants there is no guarantee that they would do well here anymore. Conditions have changed. In many places around here the vegetation is gone, the topsoil is gone, and the whole area is drier than it was before."

Another person spoke up saying that she was frustrated with the Scotch broom on her property because she could not do anything with it, and that seemed like a waste. Sensei pointed out that Scotch broom is a plant that fixes nitrogen, provides nectar for bees and other pollinators, and is ideal for revitalizing soil that has become depleted. "People cut too many trees, graze too many animals, and they plow. The amount of organic matter in the soil decreases along with the number and diversity of microorganisms. Then plants like Scotch broom come in to repair the damage. You may call it an invasive species if you like, but it was not sent by God specifically to torment you.

"Nature does not notice *how* the damage was caused; it doesn't even consider it damage. Nature simply reacts, always working to establish conditions that best foster life. Scotch broom is like a bandage applied over an open wound, and that bandage was not put there by just anyone. It was put there by the source of healing itself. Once the Scotch broom has done its job other plants will come and the Scotch broom will gradually disappear.

"Many of the species people consider 'invasive' are adapted to the kind of disturbances people cause so they follow people wherever they go. For example, plowing creates bare ground. Nature covers bare ground as soon as possible with whatever is at hand. The plants that are particularly effective at covering plowed fields, like pigweed, bindweed, and lamb's quarters, are always found together with plowed-field agriculture. Scotch broom follows the disturbance caused by cutting trees, and yellow star thistle comes hand in hand with overgrazing and eroded soil. These species seem to be a nuisance, but they are the inevitable consequence of people thinking they can somehow evade the laws of nature. We should accept responsibility and be thankful that these plants are capable of repairing the damage we have caused."

The audience became still, but just for a moment. Then the exchanges continued with renewed fervor. Sensei was *clearly* enjoying such a spirited discussion and was sorry when it became too dark to continue. Seven or eight of us stayed for dinner and we continued to chat while sitting around the kitchen table. The room was illuminated by a kerosene lamp. Sensei asked about the Maidu Indians who had lived there before. He was told how they lived at relatively high elevations during the summer months and moved to the foothills during the winter following the seasonal availability of food, and was shown pictures of the beautiful coiled baskets they used for gathering seeds and storing them, and for cooking. Then Arlo showed him a box of arrowheads and other artifacts he had collected on the farm over the past couple of years. "They just keep popping up every time I plow the fields," he said.

One fellow who was there, a poet and naturalist who lived nearby, changed the subject rather abruptly when he said, "In your talk this afternoon you mentioned that when you first saw California from the airplane you were surprised to see nearly treeless hills covered with brown grasses, and that after traveling around you have concluded that people are turning California into a desert. I think you're wrong about that. California has a Mediterranean climate where almost no rain falls for six straight months and sometimes longer. The reason California seems so desert-like this time of year is because of the climate."

This was the first of about thirty times someone said that to him during his visit. Sensei answered that he was well aware of California's climate, but the evidence of human-caused desertification were everywhere. "People just don't want to take responsibility for what they are doing and find the Mediterranean climate to be a convenient scapegoat, just like they blame 'invasive' species for run-down landscapes they have caused themselves. People are in denial about the ecological disaster they have created here."

The man was not convinced and responded by saying that he was born in California and had lived there all his life. He had studied ecology at the university and had lived for many years in a cabin he built himself deep in the forest. "With all due respect, I think I know California better than you, someone who has only been here for a couple of weeks." It was getting late so we had to end the discussion, the two of them agreeing to disagree, but beginning with this trip to California, the subject of how to rehabilitate the human-caused deserts of the world became Mr. Fukuoka's passion for the rest of his life.

The Lundberg Family Farms

The next morning we had breakfast and walked around the farm for a while. Robyn and Arlo showed us the garlic fields, some mortar stones where the Maidu had ground acorns, and several *huge* erosion gullies left by the previous owners that they were trying to repair. Then we visited with the chickens and turkeys that were running around in front of the house, thanked our hosts, and traveled on.

Our next stop was the Lundberg organic rice farm, but again, we had some time so we decided to stop for a while at a picnic area overlooking the Sacramento Valley. It was a warm, clear day. We found a table in the shade of a large oak tree and looked out at the checkerboard pattern of the fields below. We could see the Marysville Buttes, "the smallest mountain range in the world," rising oddly out of the otherwise level valley floor, and the Coast Range far off in the distance. Our conversation turned to the effect people have had on shaping the landscape of California.

Originally the coastal regions of California, the Pacific Northwest, Canada, and the islands of Japan were among the most hospitable environments for human habitation. When the Europeans came to California, ". . . they all found its hills, valleys and plains filled with elk, deer, hares, rabbits, quail, and other animals fit for food; its rivers and lakes swarming with salmon, trout, and other fish, their beds and banks covered with mussels, clams, and other edible mollusca; the rocks on its sea shores crowded with seal and otter; and its forests full of trees and plants, bearing acorns, nuts, seeds and berries."[2] The grasslands that covered the Central Valley and the surrounding foothills contained both annual and perennial plants. The most important were perennial bunchgrasses. Annual wildflowers, bulbs, forbs, and shrubs provided nectar for butterflies, bees, and other pollinating insects.

Water was everywhere. Salt marshes existed in low-lying coastal areas, freshwater marshes in interior valleys and along rivers, streams, and lakes. The lower part of the San Joaquin Valley was covered by hundreds of square miles of tule lakes. There was an eleven-hundred-square-mile freshwater marsh east of San Francisco Bay at the confluence of the San Joaquin and the Sacramento Rivers. The narrowness of the Carquinez Strait, the sole outlet from the Central Valley to the sea, caused sediment to build up behind it, forming a shallow inverted delta with sloughs and low islands of peat and tule. Where the freshwater from this marsh met the salt water of San Francisco Bay it formed a brackish estuary. Marshes and brackish estuaries like these are the richest of all ecosystems, teeming with life of all kinds. Pelicans and other birds used the marshes for nesting, and tens of millions of migrating birds used them each year as overwintering grounds.

Crystalline rivers and streams flowed unimpeded out of the mountains, and vernal pools—shallow basins that collect rainwater or snowmelt in the spring and dry out slowly—dotted the lowlands. Flooding occurred annually in the valley, leaving its rich deposit of silt from the mountains. Occasionally floods would be so great that the entire Central Valley would fill with water. These large floods were all caused by the same sequence of events, a heavy accumulation of snow early in the season followed a continuous series of warm, tropical storms that would melt the snowpack. The largest flood occurred during the winter of 1861–62 when all of the low-lying areas from the Columbia River to San Diego were inundated for weeks. These mega-floods mixed sediment from the various mountain ranges, each with its own unique mineral composition, and deposited them more or less evenly across the valley floor. This is one of the reasons that the soil in the Central Valley is among the richest in the world.

Some areas in California, such as subalpine forests and the deserts in the south, were lightly occupied and little changed by the Indians, but most of the landscape, especially the vegetation, was dramatically affected by their presence. With the arrival of the Europeans, however, the environment they had tended so carefully underwent a complete transformation as agriculture and grazing replaced native ecosystems.

The grasslands originally consisted of perennial bunchgrasses, with some annual grasses, wildflowers, and clover. The perennial grasses have deep and extensive root systems and stay green all summer. When the Spanish introduced grazing cattle, horses, sheep, and goats in the late 1700s, they also brought the seeds of European annual grasses such as rye and oats, which rapidly spread. The grazing animals selectively ate the more nutritious native perennials, giving the annuals a big reproductive edge. They also grazed the native annual flowering plants before they had a chance to produce seeds. The native grass ecosystem was soon supplanted by the annuals, causing a decrease in the organic matter in the soil, reducing the soil's ability to hold water. Grazing sheep, which naturalist John Muir famously referred to as "hooved locusts," devastated the forest meadows.

Extensive logging, especially near settlements and mining operations, accelerated during the Gold Rush, leaving entire mountainsides bare and open to erosion. Agriculture depleted the soil, turned it saline, and lowered the water table. It also eliminated habitat for pollinating insects and other wildlife. Ranchers systematically killed bears, wolves, and mountain lions by trapping them, hunting them down, and using strychnine. Birds and waterfowl were routinely poisoned by farmers and orchardists. Other animals were killed for sport or for their pelts.

Dams were built on all the major rivers around the valley. They controlled the large floods but deprived the soil of rejuvenating deposits of silt. Instead the silt accumulated behind the dams and became a problem. The tule marshes were drained to create fields for agriculture. Eventually nearly all of the wetlands were reclaimed with extensive levee systems, leaving bypasses to carry excess floodwaters to the sea. More than 95 percent of the historical tidal marsh area at the confluence of the Sacramento and San Joaquin Rivers was leveed and filled. The flow of fresh water from these rivers was further reduced by diversions for agriculture. Once the newly created fields in the delta were drained, the peat soils were exposed to oxidation. Wind erosion and decomposition of the peat led to dramatic subsidence. Today most of those fields are below sea level. Every once in a while the soil catches fire and continues to smolder for years.

These changes—the logging, grazing, alteration of grass species, reduction of wildlife, elimination of diverse habitats, plowing, dams, and marsh draining—all had the effect of making California drier. California is a much less hospitable place now than it was 250 years ago.

After we talked about these things for nearly an hour, Sensei was quiet for a moment, then a familiar sparkle appeared in his eyes. "It sounds like you don't agree with the fellow last night who said California looks the way it does because of the Mediterranean climate." We laughed, happy to be alive, got into the car and drove to our next destination.

The Lundberg Family Farms is a large, third-generation farm that was founded in 1937 by Albert and Frances Lundberg after they migrated to the Sacramento Valley from Nebraska during the Great Depression. According to the company history, "Albert had seen the ravages of the Dustbowl that resulted from poor soil management and shortsighted farming techniques. After moving the family to Northern California, he impressed upon us the need to care for the land." Albert and Frances had four sons, Eldon, Wendell, Harlan, and Homer. They all became farmers and were co-owners when Sensei came to visit, first in 1979, and again in 1986. The Lundbergs have always been leaders in the organic and sustainable farming movements and have pioneered a number of innovative, environmentally sound rice-growing techniques.

The soil near Richvale, where the farm is located, is heavy clay, which is poorly suited for many crops but ideal for rice. What began as forty acres is now sixteen thousand, almost all of it in organically grown rice. The irrigation water comes from the Feather River, which is obstructed by the nearby Oroville Dam, the tallest dam in the United States. It is an earthfill embankment dam constructed mainly from hydraulic mining debris that washed down the Feather River during the Gold Rush.

Richvale is very close to Oroville where Herman and Cornelia Aihara lived and had their school of macrobiotics. One of their students, a fellow named Peter, worked at the farm and gave a copy of *The One-Straw Revolution* to one of the four brothers. Within a few weeks they had all read it and invited Sensei to visit while he was in California. They welcomed us, and after a brief conversation we went out to the fields with Harlan as our guide.

The Lundbergs have always had a strong commitment to maintaining the health of the soil and the environment. This was reflected in their management techniques, which included water conservation, crop rotation, cover-cropping, fallowing fields, and incorporating all of the rice straw into the soil instead of burning it. Many of the fields are covered with straw after the harvest and flooded to simulate wetlands. This provides a haven for shorebirds, herons, and egrets. Other fields remain in cover crops, which also provides habitat for migrating ducks and geese. Every spring before preparing these fields for planting, volunteers, including elementary school students, carefully recover hundreds and often thousands of eggs from birds that nested in those fields. The eggs are incubated at a local hatchery and the young birds are released into the wild.

Mr. Lundberg explained all of this to Sensei, and then told him of the challenges they faced growing rice organically on such a large scale. These included weed control, the need to fallow and cover-crop fields to maintain fertility, and the need to drive heavy equipment over the soil. He asked Sensei if he had any ideas about those things. Over the course of the afternoon, Sensei did offer some suggestions on how they might modify their crop rotations as a strategy for controlling weeds.

He saw that they had fields where rice was grown one year and wheat the next, but they fallowed the fields over the winter. During the growing season they used tractors to cultivate the soil five or six times, for weed control. Sensei suggested that they do what he did and grow two crops each year, rice in the summer and wheat in the winter. Having a crop in the fields all the time would control the weeds and they would only have to make two passes over the fields with a tractor to harvest.

Here is what he suggested: "In the late summer or early fall, sow the seeds of the winter grain by airplane into the standing crop of rice, let that grow over the winter, and in the late spring, just before the winter grain is ready to be harvested, make another pass with an airplane and drop the rice seed onto the field. Then come through with a combine in May or June and harvest the wheat, leaving the stalks in the field. The rice will only be a few inches tall then and won't be hurt by the tractor. A roller attached to the combine will knock down the stalks of wheat, which will then act as mulch. The rice shouldn't have any trouble getting through the mulch and it will have a head start on the weeds. For this rotation to succeed, however, it is important that you don't plow the soil.

"The main problem is that the fields are so large and there are so few people. With this situation I don't see how you can avoid using airplanes and large machinery. You just need to make as few passes over the fields as possible." Mr. Lundberg could feel Sensei's frustration about the lack of human presence because it was also his own. Then the conversation turned to the problems of industrial-scale farming and distribution, and the high price society pays for it.

Mr. Lundberg said that very few organic farms were as large as theirs, and that growing crops responsibly, as they did, put them at a disadvantage. It is much easier and less expensive to use chemical fertilizer and burn the rice straw as conventional farmers did. The Lundbergs also have heavier crop losses from insects, but they are committed to avoiding synthetic pesticides. He also mentioned that chemical farmers pass a lot of their costs on to the environment and the public in the form of pollution. "We do have the satisfaction, though, of knowing we are producing wholesome food, and that we are doing all we can to be good stewards of the land."

I learned a lot that day, about rice growing and about humanity. These two men, from different cultures and using vastly different methods of farming, were speaking together with such respect and mutual admiration. What united them was their love of the land and their desire to leave the world a better place than they had found it.

Berkeley and the Green Gulch Zen Center

Our hosts in the San Francisco Bay Area, once again, were Mino and Fusako. Mino left the United States for Canada in 1968 to avoid the draft due to issues of personal conscience (he was later pardoned). He met Fusako in British Columbia in 1970 when she and her three daughters visited the back-to-the-land commune where Mino was living. They spent three years together in the 1970s at the Lama Foundation, a spiritual community and retreat center in New Mexico. At the time of our visit, Mino had a small construction company in Berkeley, while Fusako actively worked on projects that promoted social justice and world peace. Sensei and Ayako-san were glad to have a few days of relative quiet and appreciated the time they spent sipping tea with Fusako and her Japanese friends.

On the first afternoon, Sensei asked to see their backyard vegetable garden. Mino and Fusako looked at each other apprehensively, then Fusako said they would rather not show it to him because they were too embarrassed by its untidy condition. That refusal did not work for very long. When a sensei requests something, you pretty much *have* to comply, and soon we were walking down the back steps of the house to the garden. It felt vibrant and the plants all had good color, but it *was* overgrown. Mino explained that he had hoped to have the garden cleaned up before Sensei arrived, but he had been too busy at work.

Sensei made his way through the weeds and clover, finding a few vegetables here and there. Then he got a big smile on his face and announced that this was the most magnificent vegetable garden he had visited since coming to America. He said that the other gardens he had seen were just bare ground with the plants lined up in rows like little soldiers. But in *this* garden he could feel nature's participation. "This is the first stage of a family garden based on the natural farming method," he said. "In the first year people sow the seeds, in the second year nature makes adjustments, and in the third year the gods make a natural garden for us." Sensei even brought over small group of people the next day and held an impromptu workshop there.

One of the benefits of arranging the itinerary was that I could bring Sensei and Ayako-san to places that were of special interest to me, so one day I took them on a tour of Berkeley beginning with the campus. We entered through the North Gate, walked down the hill through a wooded area where some coast redwood trees were growing, crossed Strawberry Creek on a narrow footbridge, and went past the chancellor's house before arriving at Hilgard Hall. That building, which housed the Departments of Soil Science, Plant Nutrition, and Plant Pathology, is one of many designed by John Galen Howard using the neoclassic style of the Beaux-Arts school. These buildings, built in the early 1900s, form what is considered the "classical core" of the campus. They include, among many others, the Main Library, the Greek Theatre, Sather Gate, Wheeler Hall, Le Conte Hall, the fabulous Hearst Mining Building, and the campus's best known landmark, the 307-foot Sather Tower, better known as the Campanile, which is patterned after St. Mark's Campanile in Venice. We went inside Hilgard Hall and I showed them the classrooms and soils labs; later we visited the plant nutrition greenhouses, which were a block away. We also visited and chatted with some of my former professors who happened to be around during the summer.

We went to the top of the Campanile where we had a grand view of the entire Bay Area and the Golden Gate Bridge. Then we walked to Sproul Plaza and sat for a while taking in the college atmosphere. I told them about the long tradition of independent thinking at Berkeley dating back to the turn of the century, the Free Speech Movement of the early 1960s, and the many raucous demonstrations against the Vietnam War that were held in Sproul Plaza and often spilled out into the streets.

We walked on Telegraph Avenue, the main street leading to the campus, with its street vendors and carnival-like atmosphere, stopping at Cody's Books to see if they were selling the then recently published *The One-Straw Revolution.* They were.

Next we drove to the University Botanic Garden in Strawberry Canyon. I thought Sensei might be interested in their display of native California plants. He was—kind of. Sensei glanced at the plants for a few minutes, and then his attention turned to the weeds that were growing in the gravel pathways. I was beginning to see a pattern. He was far less interested in people's impression of nature than he was with nature's actual expression, even in a botanic garden.

Our final stop was at the Farallon Institute's Integral Urban House in west Berkeley, where four young people lived in a modest home demonstrating techniques for urban homesteading and appropriate technology. It was one of the first of its kind in the United States. They had vegetables growing in the backyard using drip irrigation, beehives, chickens, wind power, solar power, a self-composting toilet, and fifty-five-gallon drums of water inside the south-facing windows for passive solar heating. The people living there were friends from my days at the university so that made it even more enjoyable for me.

After completing the tour and thanking everyone we headed back to Mino and Fusako's house. On the way I asked Sensei what he thought of the place. He said, "I loved seeing the bright look in the eyes of the young people and their commitment to doing the best they can while living in the city. They were growing their own food, which is commendable. With so many living in cities these days it is very important for people to grow food at home if they possibly can." He reminded me that in traditional Japan, anyone who had even a small bit of land used it for growing fruits and vegetables, and most of the street trees in the cities also produced food. When I asked him about the appropriate technology demonstrations he seemed only mildly interested.

The next day we drove across the Golden Gate Bridge to the San Francisco Zen Center's Green Gulch Farm. The 115-acre farm is located about fifteen miles north of San Francisco near Muir Beach and is an inholding within the Golden Gate National Recreation Area. The farm was purchased in 1972 from George Wheelwright with the stipulations that it would always remain open to the public and would remain a working farm. The people who lived there studied Soto Zen Buddhism and sat meditation twice each day along with their other responsibilities in the garden, farm, or kitchen. The produce from the seven-acre organic farm provided for the needs of the community, and supplied the Zen Center's restaurant, Greens, in San Francisco. The rest was sold at farmers markets and at local organic food stores. There was also a one-and-a-half-acre fruit, herb, and flower garden that was influenced by the biodynamic philosophy and techniques of horticulturist Alan Chadwick, who was a student of Rudolph Steiner.

We were greeted by Wendy Johnson, the manager of the garden, and several others. Wendy took us on a brief tour, showing us the main buildings, the garden, the barn that had been remodeled into a combination student dormitory and meditation hall, and the teahouse. The architecture of the buildings was rustic Japanese. There was a pond with cattails and Japanese iris, and groves of bamboo, but there was also manzanita, coyote brush, and eucalyptus, which were typical of the surrounding vegetation. We could see the Pacific Ocean in the distance. Sensei was impressed and felt extremely comfortable there. He said that it was like a marriage of East and West, and of farming and religion. Then the Fukuokas were shown to a simple guest suite where they would stay while we were there. A few hours later we met for a delicious vegetarian dinner and went upstairs where Sensei was scheduled to speak.

The room was filled with bright-eyed Americans and Japanese, some in monk's attire, others in jeans and flannel shirts. It was an eclectic group, to be sure. I looked around and sensed trouble—not for Sensei, for me. When he began to speak my fears were confirmed almost immediately when native Japanese speakers in the audience started correcting my translation. After a few more minutes the interruptions became so frequent that we were all laughing uncontrollably. I gladly relinquished my duties to a Japanese woman who was in the audience, and was relegated to backup for farming terminology only. While listening, I realized that the reason I was having so much trouble was because Sensei had tailored his talk for his audience. He was emphasizing spiritual topics such as perception, mindfulness, the Four Noble Truths, and the Eightfold Path to Enlightenment, and that involved special religious terminology. Since I had not formally studied Buddhism, it was beyond my capability. After the talk ended, Wendy invited Sensei and Ayako-san to join them for predawn meditation, but they politely declined.

Sensei had an interesting relationship with religion. He grew up a Buddhist, but Christianity had already reached his part of Japan long before he was born. He was tolerant of Christianity and accustomed to seeing Christian symbols included in household shrines in his village. Later he would send two of his four daughters to missionary schools. He often said that natural farming was not associated with any particular religion, that it went right to the heart without need of further interpretation, but he had deep respect for people of sincere faith, whatever their path. His writing is filled with references to Christian, Buddhist, Taoist, and Hindu spirituality. "Many people have found their way to true understanding through their religious practice. My path is simply farming each day and living with gratitude. I believe that farming exists to serve and approach God." He did not meditate formally and had no daily spiritual practice other than farming.

The next morning, after a typically satisfying Green Gulch breakfast, we again walked into the garden, but this time we continued on to the fields. It was midsummer so most of the fields were filled with vegetables. Others were being prepared for planting or had a cover crop. We talked with the farm manager for a while and then continued down the path to Muir Beach about a half mile away. It was a warm, sunny morning with a steady ocean breeze. Waves crashed on the rocks and pelicans cruised low over the water. There were a few people at the water's edge, dogs vainly chasing after seabirds, and children playing with buckets in the sand. One woman was flying a kite. It was nice to have a break and just listen to the sound of the waves.

Upon our return, we were introduced to Harry Roberts, a member of the Yurok tribe. Harry went to school in the Bay Area, but spent his summers with the Yurok at the mouth of the Klamath River. There he was trained to be the "high man" of his tribe, the one entrusted with carrying the lineage of his people. During his eventful life he had been a welder, blacksmith, cook, cowboy, horticulturist, naturalist, spiritual adviser, and more. He was currently acting as farm adviser at Green Gulch. Despite his slightly curmudgeonly persona, he was respected, admired, and loved by all. Harry was a big man and was not well physically. He walked slowly with the aid of crutches. He lived nearby at the private home of one of the Zen Center elders and visited the farm as often as he could manage.

When Sensei and Harry met they recognized immediately that they shared a deep spiritual bond. Harry said, "It is so good to meet someone I can really communicate with. I have been pretty lonely for a long while." Sensei, for once, was speechless. They became instant friends, and, after getting to know each other during the course of the afternoon, realized that they were also fellow warriors who cared passionately about protecting the earth. Later Sensei said to him, "You must be the guardian deity of the redwood forests."

"That's right. Say, you just said something very nice there," Harry answered with a grin.

We decided to visit Muir Woods National Monument just a few miles away where there is an impressive stand of old-growth coast redwood trees. The path through the park is level and well maintained, so Harry was able to negotiate it, although with some difficulty. We took the short half-mile loop through the forest but even that took us quite a while with frequent stops to rest and admire the forest. Besides the towering redwoods there were big-leaf maple, Douglas fir, and tanoak trees, sorrel, sword ferns, and huckleberries. The two of them talked about the vegetation and how similar it was to the original forests of Japan.

Then the conversation turned to plant diseases. They agreed that almost all of them originate in the root system when the plant is under stress. Stress is often caused by a change in environmental conditions that affects the mycorrhizal association between the roots and the fungi living in the soil. Sensei told Harry about a disease of pine trees in Japan that began that way, but instead of treating the source of the problem, the Japanese Forestry Department had sprayed the forests with pesticides to kill the beetles that were only finishing off the infected trees. Harry said with a smile that it sounded like something our own forest service might do.

Then they talked about the condition of the landscape in California. Sensei told him that he had the impression people were thoughtlessly turning California into a desert, but no one seemed to notice or care. Harry agreed. He told Sensei about his early life on the Klamath and how during his lifetime he had witnessed the widespread destruction of the redwood forests.* He believed it was crucial that those areas be reforested. However, even though redwoods grow to be the tallest trees on earth, their root system is very shallow. Since so much erosion had occurred over the past 150 years, Harry was not certain redwood trees would thrive anymore, even if they were replanted where they had grown before.

Sensei had an idea. In Japan there is a species of *Cryptomeria*, or Japanese cedar, that is closely related to redwoods but has a deeper, more robust root system. It might be able to survive on the eroded slopes by penetrating the sandstone bedrock of the Coast Range, building soil as it grew. Harry knew the species, and they agreed to try the experiment. After he returned to Japan, Sensei sent a student to a national forest reserve that protected one of Japan's last remaining old-growth forests, where he collected seeds from that species of tree. Sensei knew of the place because it is in Kochi Prefecture, near the agricultural experiment station where he worked during the war. He sent the seeds to Harry. In return, Harry sent Sensei a cup that had been hand-carved from the branch of a redwood tree.

Harry sowed the seeds in propagation trays with the help of the Green Gulch students. Later they transplanted the seedlings into small pots. When they grew large enough, they were planted at the Green Gulch Farm in the three areas Harry had suggested.

On our way back, we stopped at a turnout where Harry pointed out a rock outcropping on the hill where he sat when he was a young man. "At the time, my feet reached the ground, but so much soil has eroded since then that now my feet would just be dangling in space," he said. Harry was sitting on the lowered tailgate of his yellow pickup truck with Sensei at his side. The rest of us were gathered around. "So how are we going to put this sorry world back together, Harry?" Sensei asked. "It's not the ideal situation, but I guess we'll have to count on these people," he said good-naturedly, referring all of us. Then one of the students suggested that Harry and Sensei were like two old-growth

* Today only 4 percent of the nearly two million acres of original old-growth redwood forests remain.

redwood trees and we were all just sprouts from the stump.* Everyone got a good laugh out of that.

I visited Harry many times over the next year and a half. Mainly, we talked about his people and their relationship with the earth. He told me about his life, his native plants nursery on the Russian River, and his time working and informally studying at Berkeley. In the 1950s he and two others installed the native California plants section at the UC Botanic Garden that Sensei and I had visited. I told him how Sensei was mainly interested in the weeds growing in the walkways, and that amused him.

Sometimes, but rarely, we talked about spiritual matters. He seemed at peace with himself and his place in the world. "Many people say they love nature," he said, "but when you calm your mind, and are true to yourself and others, you can feel the earth loving you back."

Mr. Fukuoka returned to the Green Gulch Farm seven years later. Wendy and some other students showed him a photo of Harry carefully planting the *Cryptomeria* seeds he had sent into a seedling flat, and then took him to where the trees were growing. Some were already five or six feet tall. They said that sowing the seeds of those trees was one of the last things Harry did before he passed away in 1981. Then they pointed across the valley to a simple hillside stupa** where his ashes were buried.

Sacramento

Our next stop was in Davis, where Sensei met with students from the university's organic farm. In the evening he gave a talk at the site of a seventy-acre, 220-home subdivision known as Village Homes. It was designed to encourage a sense of community as well as conserve energy and natural resources. The layout is compact with narrow streets and bike paths. The network of creek beds, swales, and ponds allows

* Redwood trees send new shoots from their base even while they are still standing. After they are cut or are damaged by lightning, new shoots sprout from the stump becoming "second-growth" trees.

** A Buddhist commemorative monument that contains sacred relics.

rainwater to be absorbed by the soil, the buildings are oriented to take full advantage of passive solar energy, and most of the houses have solar panels on their roofs. There are edible trees, shrubs, and vines in many of the common areas, twenty-three acres of orchards, two large community garden plots, as well as personal gardens. Residents are free to harvest from the common areas, but only as much as the family can consume itself. Today, Village Homes is known worldwide as a successful model for this type of community.

When Mr. Fukuoka visited in 1979, however, only some of the buildings had been completed and the landscaping was about to be installed. The drainage contours, ponds, and general layout were already in place so it was possible to get a good idea of what the community would be like. In his talk, Sensei revisited the themes of how important it is for people to produce their own food and how that goal would be facilitated by a redistribution of the population to rural areas. He loved the idea of having commonly owned orchards, berries, vines, and edible shrubs lining the streets.

He said that he thought many of the world's problems would disappear if there was food growing everywhere and available to all. "In the Middle East, for example, people are continuously in conflict. I realize that religious views play a role, but if fruit trees and other food-producing plants were growing all around as they once were, I believe that a lot of the hostility would end." As he was concluding, the sun dipped behind the mountains and a cool breeze blew in from the delta. That was a relief since it had been over one hundred degrees that day.

Whenever possible, we traveled on back roads. Our short trip from Davis to Sacramento took us through field after field of tomatoes, which seemed appropriate since the locals sometimes refer to the state capital as "Sack-o'-tomatoes." At one point we found ourselves behind a large truck piled high with, what else, tomatoes. When the truck went over railroad tracks, the load was jostled and some of them fell onto the road, but instead of going *splat*, as you would expect, they bounced. Sensei was surprised. "What fruit is that?" he asked. "I thought they were tomatoes." "They are, but when you hold one in your hand it feels like it's made of rubber," I replied. We stopped and picked one up so he could see that they were hard as tennis balls. "They are a special variety of Roma tomato used for canning," I said. "The agriculture department at the University of

California in Davis was given a grant by agribusiness to develop a tomato that would not bruise while it was being shipped. This is the result of their work, a tomato that bounces when it hits the ground." When I was studying soils at Berkeley, we mockingly referred to that variety as "the rubber tomato," and to the agriculture department at Davis in general as "the land of the rubber tomato" because of all the strange creations that were developed there at the behest of big agriculture.

I was working for the California State Department of Forestry at the time of his visit and had arranged for Sensei to speak at a brown-bag luncheon in the Resources Building. About forty people representing a cross section of the various resources agencies were there. He was pleased to have the opportunity to discuss land-use issues with them. The talk was sponsored by the Department of Conservation, which is one of six departments within the Resources Agency. The others are Water Resources, Fish and Wildlife, Parks and Recreation, Forestry, and the Conservation Corps. Before he spoke, we were invited to meet with the department head, a geologist named Priscilla Grew. Ms. Grew had already read *The One-Straw Revolution* and was pleased to have a private conversation with Mr. Fukuoka.

Sensei began by saying that he had noticed that the vegetation of California and Japan were quite similar. "Perhaps there are clues from the geology that might help me understand why that is." Ms. Grew went to her library and returned with a book that had a geologic map of the Pacific Region. She showed him that the parent rock of Siberia and Alaska were the same; Hokkaido and southern Canada also corresponded, as did the parent minerals of central Japan and California, and Southeast Asia and Mexico. Even Mount Fuji and Mount Shasta, which are both considered to be sacred mountains, are in corresponding positions. She suggested that the two continents were part of the same landmass at one time.

Then the conversation took on a more personal tone. Ms. Grew asked Sensei about his experiences in Japan, and how the tour was going so far. I had known her for some time so I was not surprised when she winked at me and asked Sensei if I was taking good care of him and his wife, and if not, he should not hesitate to give her a call. He laughed and thanked her. Sensei then asked Ms. Grew how she had become head of such an important government agency. Wasn't it unusual for a woman to be in a position of so much authority? She said that the governor, Jerry Brown, was trying to do things differently, and that included making appointments that some considered unusual. But it was not only in state government; the norms of society were changing, and many traditional ways of thinking were being challenged and discarded. The result was that women now had more opportunities than they did before.*

After the talk, we took the elevator back to the lobby. The Resources Building was just a few blocks from the state capital, so we decided to walk there. The park that surrounds the capital building is actually more of a botanical garden than it is a city park with its large collection of trees from all over the world. When we passed an area that was filled with colorful bedding plants I asked him if he was interested in plants like that. He waved his hand disinterestedly and said, "No, not really."

Sensei wanted to see where I was living so I brought him to the converted garage I called home. It wasn't much, like living in a cabin, really, but perfectly sufficient for my needs at the time. Sacramento is crisscrossed by freeways and surrounded by uninteresting suburbs, but the old section of the city, where I lived, was quite nice with ornate Victorian houses and tree-lined streets. The house next door had a small synthetic lawn in front, and as we were leaving, we saw the homeowner clearing leaves from it using a leaf blower. I thought that bizarre sight would likely trigger a rant from Sensei about "artificial green," which is what he called lawns, even those with real grass, or a lament about how far humanity had alienated itself from nature, but apparently the sight was too surreal even for him to fully process it. He smiled, shook his head, and said, "Let's just pretend we didn't see that, all right?"

The East Coast and Los Angeles

Mr. Fukuoka had been invited to visit Rodale Press, and to teach at the East West Foundation's macrobiotic Summer Conference in Amherst, Massachusetts, so we boarded a plane and flew to the East Coast. Our first stop was New York City.

---

* While I worked in Sacramento I met another recent graduate from Berkeley who was the first woman to become a licensed forester in the state of California.

After all he had heard about New York, Sensei was glad to have time to explore the city. We took taxis, rode the subway, and spent hours wandering around on the streets. A Japanese woman named Mogi-san who had stayed at his farm was living on the Lower East Side so we visited her, then walked together to Greenwich Village, the garment district, Chinatown, and even visited the flagship store of the toy seller FAO Schwarz on Fifth Avenue. At Sensei's request, we also went to Harlem. He enjoyed the ethnic character of the city and that special energy that can only be experienced in Manhattan.

He had heard that New York City was a dangerous place, but he felt perfectly comfortable there. Here is what he wrote about his visit:

> *I . . . spent several days in New York and even walked about some at night. I found none of the individuals that I met frightful or menacing, whether in Harlem or elsewhere. They all seemed to be very good people. I even thought that, if anything, it was the blacks who were able to laugh from the heart . . . But when I looked at the faces of the smart and clever, those living affluent lives, none of them bore an expression of contentment. All had a tragic, cornered look on their faces.*[3]

Our next stop was Rodale Press, in Emmaus, Pennsylvania. Sensei was looking forward to this visit to "the inspirational and educational leaders" of the organic farming movement. They had published *The One-Straw Revolution* the year before and he wanted to thank them. He also hoped to discuss no-tillage techniques and strategies for slowing, perhaps reversing, the troubling developments in society and modern agriculture. Carol Stoner, in-house editor for *The One-Straw Revolution*, and Jeff Cox, managing editor of *Organic Gardening* magazine, joined Sensei, his wife, and me for lunch with Robert Rodale. The vegetarian meal was prepared right there at the Rodale Test Kitchen.

The conversation was very pleasant. Mr. Rodale asked Sensei about the various varieties of tomatoes and other vegetables he grew in Japan, and then went on about the ones they had found promising at their research farm. Then they discussed more gardening techniques. Finally, when the luncheon was nearly over, Sensei changed the subject abruptly to no-tillage farming, making the point that when a permanent ground cover of white clover is grown, there is no need to go to the trouble of making compost. Making compost is a lot of work, and he did not particularly like doing unnecessary work. Also, he said that farmers generally have difficulty transporting and spreading compost over very large fields. Mr. Rodale suggested that we visit their three-hundred-acre research farm in Kutztown to see the no-till trials they were doing there. So we did.

Dr. Richard Harwood, the director of the farm at the time, showed us around and explained the various experiments they were doing. One was the "amaranth project," which was designed to determine whether or not it was feasible to use amaranth as an everyday grain in the United States. Sensei was most interested in the plots where they were testing vegetable production in fields that had not been plowed and were planted with white clover. That experiment, Dr. Harwood said, was inspired by Mr. Fukuoka's example. Sensei thanked him and offered a few tips.

As we continued on, we passed two *very* large mounds of compost. That got Sensei going again on the topic of how impractical it was to use compost on large acreages and how much easier it is to just toss out seeds and grow green manure. This was vintage Sensei being Sensei. There we were, at the headquarters of Rodale Press and the Rodale Research Farm, the vanguard of the worldwide organic farming movement, and Sensei was spending most of the time questioning the need to make compost. I suppose that when you're a sensei you can get away with that sort of thing. I heard that when he was in Europe speaking to a group who primarily raised livestock, he referred to Europe as "one big overgrazed and eroded cattle ranch." And when he was in India he took them to task for all the sacred cows everywhere.

The founders of the East West Foundation (now the Kushi Institute) were Michio and Aveline Kushi. They were also students of George Ohsawa in Japan. Like the Aiharas, the Kushis came to the United States to introduce macrobiotics to the West, but their approach and personal styles could not have been more different. The Kushis were much more formal than the Aiharas, and they were very much about business. Besides teaching, they established the Erewhon Trading Co., opened one of the first natural food stores in the Boston area, and published a magazine, the *East West Journal*, which had a large worldwide circulation.

The Summer Conference was held at Amherst College in western Massachusetts. The campus was landscaped with spacious lawns, flowering ornamental shrubs, and stately trees. Students stayed in dormitories, meals were cooked in the university kitchen, and most of the classes were held in lecture halls. This was a far cry from the French Meadows camp where everyone slept under the stars and the classes were held outdoors with the students sitting cross-legged on the ground. Still, the setting seemed appropriate for the nature of the conference. Sensei and Ayako-san stayed in a private suite and enjoyed interacting with the students very much.

It was a macrobiotic program, so his talks mainly concerned diet and how diet affects clarity of thought, but in one lecture he asked the students to consider the forests of New England. At first glance they seem to be a "sea of green" and quite natural, but he noticed that the soil was poor and seemed to be worn out. Some people told him that was the result of earlier glaciation, but he suspected that the land was cleared for grazing at one time. When the soil became eroded and unproductive it was abandoned for richer soils elsewhere. He suggested that the forests we see today are merely a shadow of what they had been long ago.

"Everywhere we look we see only ourselves, not true nature," he said. "Take the grounds of this university, for example. It looks lovely with its expansive lawns and large trees, but what you are seeing is people's aesthetic impression of nature, a pale imitation of the real thing. It reflects the character of the university with its sense of order and self-importance and is designed with human pleasure and convenience in mind. True nature has been banished from this campus. The lawns are pleasant to look and to sit on, but I have not seen any butterflies or insects there. I call lawns 'artificial green.' As far as nature is concerned, they might as well be concrete."

One afternoon between lectures, Sensei did an interview for *The Mother Earth News*. I had called them a few weeks earlier suggesting the interview, but they had no one available to come to Amherst. We agreed that I would do the interview for them, but I did not mention that to Sensei. At the appointed time I said, "All right, it's two o'clock, let's get started." "But the reporter hasn't arrived yet." "*I'm* the reporter," I replied. At first he scowled at me. He knew I had pulled one over on him and apprentices were not supposed to do that sort of thing. Then he smiled to himself, as if realizing the beauty of the situation. He took out his drawing pad, ink, and brushes, and drew a beautiful mountain scene with a poem to the side. "Okay," he said, "here's how we'll start. You ask me what I am drawing. I'll answer and recite the poem. Then you ask me to explain the poem. After I do that, you can ask me anything you want."*

Our flight from Boston to Los Angeles, the last stop on the itinerary, took the southern route over Texas, New Mexico, Arizona, and the deserts of Southern California. After more than two hours of nothing but arid lands, the sprawling Los Angeles metropolitan area came into view. This was a real shock to him. He asked, "How many people live here?"

"Oh, about twelve million," I replied.

My parents, Irving and Vivian, greeted us at the airport. Instead of taking the freeway home, we decided to travel on city streets. That route took us through Beverly Hills, one of the wealthiest areas of the city, with its broad, tree-lined avenues. The houses are all similar: A large lawn bisected by a central walkway leads to the front door of a mansion. Ferns, semi-tropical plants, and colorful annual flowers grow in beds in front of the house and along the sides of the lawn. Sensei was horrified as we passed house after house like that. He said, "Twelve million people come to live in the desert and they landscape with lawns, ferns, annual flowers, and tropical plants, all plants that need *extra* water. It's appalling!" Ayako-san, who was looking at the same scene through the other back window, said, "Twelve million people come to live in the desert, and look what pleasant surroundings they have created for themselves. It's fantastic!"

We had no scheduled appearances. The Fukuokas enjoyed the two quiet days getting their things together and talking casually with my parents. On one of those mornings, Sensei brought me out to the sidewalk in front of the house. "Look around," he said, "nothing but

---

* The interview was published in *The Mother Earth News* 76, July–August 1982. It is reprinted in appendix C.

concrete, lawns, exotic ornamentals, and buildings. It must be hard for people living here to maintain a positive attitude." I told him that I had grown up in that house and hadn't really noticed. It just felt like a normal neighborhood to me. We spent the afternoon hiking in Griffith Park near the HOLLYWOOD sign so he could get an idea of the topography and the vegetation that existed before there were so many people living there. In the evening we had an informal get-together with friends and some neighbors.

The next day we reminisced about some of the more enjoyable moments of the trip as we sat in the airport waiting for the departure of their flight back to Japan. Sensei said, "Larry, thank you above all for bringing me to Los Angeles."

That surprised me. "We went to so many beautiful places and did so many interesting things during your visit. Why would you say that about Los Angeles?"

He said, "Remember that young woman we talked with at the party last night, the woman who worked for the magazine? I asked her how she felt about living in a desert where it rarely rained, and she said, 'I hate it when it rains. It's so inconvenient!' This was very important for my research. I would never have *dreamed* that people could be so disconnected from the place where they lived."

When the announcement for boarding was broadcast, Sensei thanked me and said, "Next time, schedule a few events where not everybody agrees with me, okay?"

Q and A at the Second International Permaculture Conference

On his second visit to the United States in 1986, Mr. Fukuoka attended a meeting for permaculture designers at Breitenbush Hot Springs in Oregon. This was followed by the 2nd International Permaculture Conference, which was held at The Evergreen State College, in Olympia, Washington. At Breitenbush he enjoyed soaking in the natural hot springs, hiking in the towering forests of the Cascade Mountains, and chatting with the other one hundred or so attendees who had come from all over the world for the two events. He met Bill Mollison, the co-founder of permaculture, and Allan Savory, the creator of "holistic resource management" in which domestic livestock are used like herds of wild animals to reverse desertification and rehabilitate deserts and damaged rangeland. Sensei also gave a few talks and held an impromptu seminar at the site of an abandoned vegetable garden.

The conference in Olympia was open to the public, and more than 750 people attended. There were a number of talks and workshops during the two-day event, as well as a delightful three-way discussion in the main hall among Bill Mollison, Wes Jackson, the founder of The Land Institute in Salina, Kansas, and Mr. Fukuoka, which he describes in *Sowing Seeds in the Desert*. Mr. Fukuoka also held a question-and-answer session that touched on a wide range of the topics. He began by making a short statement, then opened the floor to questions:

"When I started farming forty years ago I had one of the smallest farms in a poor village in southern Japan. My father had owned a large farm, but after the Second World War the land was redistributed, so I began with very little land and very little money. I took advantage of a government program that allowed me to plant trees and I was able to buy some poor orchard land very cheaply. Now I have a productive citrus orchard with thousands of trees and shrubs of all types. I also grow rice, barley, fruits, berries, and vegetables. I did that by closely observing nature and finding the most appropriate plants for that area. Nature showed me what to do every step of the way.

"You can easily make the mistake of thinking that the nature we see today is the original nature of an area. For example, if you look at the grain fields of Oregon or the agricultural valleys in California it is not at all like what the land was originally. It is more of an imitation nature created by human will and human actions. To find the original nature in these areas it is useful to visit mountain areas like the Olympic Mountains here in Washington or the area around Breitenbush Hot Springs in Oregon and then imagine what the lowlands must have been like.

"When I went to Africa it was nearly impossible to see the original landscape so it was hard to decide which plants to sow and when. I was told that many years ago Somalia was as forested as the state of Washington. Ethiopia is almost entirely desert now but was once lush forests. I decided to sow the seeds of over one hundred varieties of vegetables and over one hundred varieties of fruit trees and ground cover plants after encasing them in clay pellets. That way, in just one year, I could see what trees and plants would grow well there. In deciding what trees to use, I suggest you look first to what is native to the area. Palms, papaya, and banana are accustomed to the climate there so they will grow very quickly, but citrus and pomegranate, which are not native to that part of Africa, also do well in the present conditions and should be included as well. It is a good idea to sow the seeds together and at the same time. Trees, vegetables, ground cover plants, and microorganisms, all living together with animals—that is nature's way. The idea is to regenerate the land so nature is whole and everything can work in harmony again.

"Here is another idea for regreening the deserts. The desert looks very dry, but there *is* water and there *are* rivers. The first thing to do is to create greenbelts along the banks of the rivers using acacia trees or another fast-growing, nitrogen-fixing species. If you plant these trees perpendicular to the bank and space them sixty feet apart or so, the water will filter up through roots of one tree and will be carried to the roots of the next. This creates a kind of plant-based irrigation system and the area around the trees will become moist. Then I would plant twenty or thirty different kinds of trees and sow the seeds of other shrubs and vegetables. Using plants to build the irrigation system is much more effective than building dams and irrigation canals.

"Now I would like to talk about my philosophy for a moment and then we can go on to your questions. Natural farming does not separate the soil from the plants, water, or animals. Trees, grasses, animals, micro-organisms, even the sky and the mountains are all part of one 'big life.' Everything is alive and exists as one body. I could spend an entire day talking about the characteristics of an acacia tree. I could go on about clover for an entire week, but that would not give you the slightest idea of their essence or how they fit into nature's design. Scientists believe they will eventually come to understand how nature works by studying its parts. The geneticist sees a tree from the point of view of genetic science, the soil scientist from their view of the soil, the botanist from the point of view of the biological sciences, and so forth. But the more specialized knowledge becomes, the further it moves from truth.

"A couple of months ago some university professors came to my farm and asked me to help them with a study they were doing. They wanted to plant fruit trees using the natural method, but first they needed to do a report to help them decide the best way to go about it. Hundreds of thousands of dollars had been set aside for this study, which they estimated would take twenty years. The title of the report was going to be, 'A Study of the Benefits and Techniques for Planting Fruit Trees Among Broadleaf Trees in Low-Quality Terrain.' They were planning to study the temperature, the humidity, the light, the soil conditions, and anything else they could think of. While they are studying, of course, the soil is continuing to erode and chemicals are being spread all over. The whole world might be gone by the time they are finished.

"It took me twenty years to learn how to grow plants by the natural method, but at the end of that time I had rich soil, fruit trees, tall acacias, pines, shrubs, berries, a mixed ground cover with chickens running everywhere. All those scientists needed to do was watch me as I scattered seeds. They could have learned everything they needed to know in just one year."

Q: (from the audience) You have talked a lot about using white clover to improve the soil. Why do you use white clover instead of other kinds of clover?

A: I tried many different types of clover and found white clover was the best for me. The roots form a mat in the top few inches of the soil so weed seeds coming up from below cannot get through and windblown seeds also have a hard time germinating. Red clover gets so tall that it gets in the way of the vegetables. If you look out at a field of red clover it seems to be solid just like white clover, but they are individual plants so weeds can find their way between them. Just like everything else, though, the only way to know what kind of clover will be right for the place you live is to try them and see what happens.

The best time to sow clover is in the late spring, or early fall. I only use a handful, maybe a quarter pound for a quarter acre, but you can use more if you want to. You will need to reseed the clover every four or five years to keep it strong. I am not trying to eliminate the weeds, just keep them in balance with the other ground cover plants. I have white clover growing everywhere in the orchard and the rice fields at my farm.

Q: You also talk a lot about acacia trees. We can't grow acacia trees here in Washington because it gets too cold. What should we use instead?

A: I have seen ten or twenty kinds of trees just walking around in the woods over the past few weeks that would work just as well. Except for alder and black locust I don't know their names, but I can see from their form that they would be fine. Any nitrogen-fixing tree that grows easily will do, but its roots should grow deep rather than spreading along the surface; otherwise they will interfere with the vegetables. Of course, they are still good for the soil.

Q: You harvest a lot of rice and ship so many mandarin oranges for sale. Doesn't that deplete the land?

A: Not at all! In fact the soil becomes richer every year because I have a continuous ground cover enriching the soil, lifting nutrients from deep down and depositing them on the surface. All I take is the seeds and the fruit. Everything else goes right back into the fields. It's like making compost right there in the soil, but without the work.

Everyone seems to love compost. Organic farming and traditional Japanese agriculture are based entirely on composting. People like compost because it is very fast. You can have concentrated fertilizer in a matter of weeks. If you cut the grasses and weeds and leave them on the surface, it might take five or six months for them to decompose, but this is the natural way and, in the long run, it works better. The desire to make concentrated compost so quickly is the same mentality of people who want to drive in a fast car instead of walking or riding a bicycle. Besides, with compost, you can only do it when you are young and full of strength. Anyone can toss out seeds, cut the weeds, and let them fall to the ground, even an elderly person. Those of you who like to do work just for work's sake, go ahead and make compost. I don't mind. Just as long as you know that it is not necessary.

Q: What do you do with scraps from the kitchen?

A: I put it back into the soil. *Everything* goes back, I just don't bother to make it into compost first. I bury it just below the surface. I put it here one time, and over there another.

Q: Do you do the same thing with human waste? Wouldn't that spread disease?

A: Yes, I just spread it around in the orchard, here and there. If you have pigs they will root around and find it, eat it, and then chickens and other birds will eat the pig manure and microorganisms will decompose the bird manure. It will all be gone in a matter of a few days and it doesn't smell. Nature provides a perfect system for keeping the earth clean. People think that the world outside is dirty and full of germs. It is actually the cleanest place of all. Disease problems occur when people, or cattle, or fish are crowded together in one place. Then the manure piles up, leading to pollution and disease. I consider human manure to be a gift from the Buddha.

Q: How do you begin to practice natural farming?

A: Stop thinking so much. Make your mind like the mind of a baby. They see the world directly without distinctions or judgments of any kind. When you go outside with a clear mind the trees and plants appear so beautifully. The less you think, the more beautiful things become. When you look with your eyes, you can only see the surface. When you look with your heart you can see everything. People knew this in ancient times. We have just forgotten. Nature did not forget humanity, but humanity has forgotten Nature.

Q: Do you coordinate your farming activity with the phases of the moon or the position of the planets?

A: The land and the seeds themselves know the proper timing for everything. Planting by the phases of the moon is an intellectual idea. It is something people have thought up, and therefore I don't use it.

Q: Many people trim ornamental plants and even some perennial vegetables in the fall so they will grow out more vigorously in the spring. Do you do that?

A: I would put that practice into the category of a horticultural trick. I allow plants, even the commercial citrus trees, to grow to their natural form. Trimming plants back and pruning orchard trees is really designed to make the plant bow to human will. It comes directly from the mentality that believes that nature exists solely for the use and benefit of human beings. If you follow that way of thinking the next thing you'll be saying is, "I want to grow this kind of plant" and "I want to grow that kind plant," even if it is from another area and needs special care to survive. It all begins with desire, with the thought that you want something.

Here's an example. One young Japanese fellow stayed at my farm for more than three years to learn about natural farming. Later he came to California with the idea of making the landscape green. He tried to grow rice on a dry hillside in the Coast Range where there was almost no water. He failed because he decided that he wanted to grow rice without first consulting with the spirit of the land and asking what *it* wanted. The first question you should ask when you come to a piece of land is, "What does the earth need?" not "What do I want to grow here." "How can I serve nature" instead of "What can I get from nature?"

Q: Do you grow all your trees from seed or do you graft some of them?

A: From the standpoint of natural farming growing trees from seed is best, but I also use some grafted stock for the commercially producing mandarin oranges. If you grow fruit trees from seed the quality of the fruit is rarely good enough to sell, but it helps maintain the genetic diversity of the orchard and produces some very unusual fruit in all sizes and shapes. I see new and unexpected wonders every time I walk through the orchard. It is one of the joys of living in a place where nature is free to express itself. The saplings grown from seed become magnificent trees with natural form. Most of the fruit goes unharvested, but it becomes food for birds and other animals and then food for microorganisms. I'm not growing food just for human beings, you know.

Q: I am a plant breeder from New Zealand and I spend a lot of time selecting from trees that were grown from seed. Is there any way you can select a tree specifically for the characteristics that will make it good for natural farming or sustainable agriculture?

A: If you go through the seedlings searching for certain characteristics, even if they are characteristics that would make them better suited for natural farming, you would still be consciously making the selection using your discriminating mind. It is better to plant the seeds and just enjoy watching as nature's plan unfolds.

Q: Can you practice natural farming in a backyard garden?

A: Yes you can. Natural farming is actually a state of mind. It is living and growing food for the sake of nature, for the sake of the gods. I visited a vegetable garden with a fellow named Katsu just a few days ago at Breitenbush Hot Springs. Katsu planted and tended a vegetable garden there for one year but had to abandon it for a few years while he went to live somewhere else. This was the first time he had seen the garden since he left. He was embarrassed because the garden seemed overgrown to him.

Most people would have judged that some parts of the garden looked pretty good, but other areas did not look so good. They see good plants and bad plants, good insects and bad insects; they see the weeds as separate from the vegetables and the birds as separate from the mice, gophers, and earthworms. When the gods viewed Katsu's garden they did not see any of those things. They saw everything all at once without judgment.

Actually, the garden was doing fine. Initially, the garden was created by people for the sake of human beings. Since then, nature and the gods have been working together in a kind of partnership. All the gardener needs to do is join in. The most direct way to do that is to empty your mind and proceed with a spirit of humility and gratitude. Nature *wants* to work with us if only we will allow it.

Some people have asked me if it is possible to use natural farming for growing vegetables on a commercial scale, but that is not its best use. Not that it can't be done, but it just doesn't go along with the whole idea of it. In order to make it economically viable, farmers start to think that it would be more efficient to plant in rows; then they start pruning the trees so the fruit can be harvested more easily. Pretty soon they are right back to asking, "What can I get from nature and how can I do it most efficiently." Then it is not natural farming anymore.

Q: Please tell us a little about the economics of your farm. Have you managed to make a living through farming alone?

A: I haven't become rich, but I make a decent living. I ship a thousand cases of natural mandarin oranges to Tokyo each month for about six months of the year. Each case weighs fifteen kilograms (thirty-three pounds). They are a little small and have a few blemishes on the skin, but they are the best-tasting oranges on the market. They are also the least expensive. Fruits and vegetables grown naturally should be inexpensive because it involves so little work and there is almost no overhead. The price I charge for them reflects that. I also make sure that the shop owners do not gouge customers with a high markup. That would be unfair and it would give people the impression that natural food is expensive to produce. I also sell most of the rice and barley I grow. I make an effort to sell my produce to the cities because many people there have never tasted the flavor of naturally grown produce.

Farmers in Japan try to make about fifty thousand dollars each year. That is their goal from one or two acres. Small-scale organic farmers in the United States have more land, but I'm not sure that they are making much more than the Japanese farmer, maybe even less.

Q: There are so many problems in the world today that sometimes it seems overwhelming. Where would you start to set things right again?

A: I will answer this question in the same way I answered the question about how one should start practicing natural farming: Stop thinking so much. The problems we have today—the destruction of nature, continuous warfare, overpopulation, disease, anxiety, depression—are all the result of people separating themselves from and trying to improve upon nature. That shuts them off from the source of true knowledge and true wisdom so the problems we see in the world are inevitable. People feel that they can make their way perfectly well in the world using their intellect—and look at the mess they have caused. Until people stop using their intellect as their guide, these problems will not go away; in fact, they will continue to get worse.

If we would return to that earlier state of mind and live by natural farming—that is, letting nature grow the food and maintain a healthy balance—there would be peace in the world and all of the other problems would disappear. There would be no private ownership of

> property, no boundaries between countries, and wherever one went there would be food growing. This is not a dream. It *can* actually happen, but society has to decide that this is the direction they want to go. No single person can make this happen, but anyone can make it so in their own lives.

Mr. Fukuoka turned over management of his farm to his son when he was in his mid-eighties but he continued to travel and give talks until he was too weak to continue. He passed away peacefully at his home on August 16, 2008, at the age of ninety-five. It was during the Obon festival when the ancestors come to visit the living for three days. On the evening of the third day the ancestors return with a send-off of songs, dances, and fireworks. Fukuoka-sensei died on the third night of Obon.

Although I have not returned to Japan since 1976, and did not personally meet Mr. Fukuoka again after his second visit to the United States, I continued to be touch with him and my friends in Japan. The farm at Shuzan was turned into a Buddhist study center and my friends were asked to leave. Bill, Hiroko, and their three kids moved to Santa Cruz, California, as did other friends from Kyoto and the *buzoku* community. We still get together every year at a state park in the redwoods to celebrate a traditional Japanese New Year. We make mochi rice cakes by pounding steamed glutinous rice with large wooden mallets, sing songs, sip tea, and just enjoy being together again.

After *The One-Straw Revolution* was published I held a variety of jobs, most of them related to nursery work, and traveled from place to place visiting farms and rural communities in California and the Pacific Northwest. In 1982 I worked with Tilth, a loose-knit group of organic farmers and activists in the Pacific Northwest, on a book called *The Future Is Abundant*, which was a practical guide for applying natural farming and permaculture in the region. Eventually I settled into being a parent and householder in Oakland, California. I ran a residential landscaping business then, basically remodeling people's yards. I always tried to steer my clients toward environmentally sound decisions, but still, it was a far cry from natural farming. When my daughter graduated from high school and went off to college in Colorado, I left the city for the small town in southern Oregon where I live today.

All during that time I gave talks and workshops about Mr. Fukuoka and his way of farming. There were always many questions, most of them related to the farming techniques, but some were about the philosophy that reminded me of the same questions I had asked myself while I was living on the mountain and for some time after that.

Another thing that often puzzles people is the question of where natural farming fits in with other forms of agriculture: Is it a primitive form of agriculture? An offshoot of traditional Japanese agriculture? An ecological form of organic farming? Some even consider it to be a precursor to permaculture. When I first realized the answer to that question, everything fell into place for me. It's all about culture. Natural farming is nearly identical to the way Indigenous people lived all over the world before the advent of modern civilization. I also came to understand how Mr. Fukuoka could speak so highly of traditional village farmers in Japan and still criticize their farming methods so severely. He also criticized organic farming and permaculture, which surprised me at first, but now I understood exactly why. The next three chapters of this book compare natural farming with these other forms of agriculture and their cultural origins in the hope of shedding light on what natural farming actually is.

CHAPTER 5

Indigenous Ways

Although Mr. Fukuoka grew up in an agricultural society, his outlook and practices were more closely aligned with those of Indigenous people whose traditions and way of life encompass the entire span of human history. Ten thousand years ago there were countless tribes all over the world, each occupying a specific territory. Each community was unique, but almost all of them shared a common social and economic model, and a common code of ethics that was based on sustainability. Protecting nature's ability to replenish itself and the long-term survival of the tribe were fundamental to the way these humans lived.

First Peoples World Wide, an organization that funds local development projects in Indigenous communities all over the world, suggests four central principles that were followed by the most successful tribal societies:

1. ***Community Is Essential for Survival.*** *Concern for the greater good and respect for the community are embedded in Indigenous legal, political, social and economic structures. Equity and justice are essential to the sustainability of the Indigenous model, freeing the community to cooperate rather than forcing individuals to compete for scarce resources.*
2. ***Life Is Sustained Through Balance and Harmony.*** *The effectiveness of traditional Indigenous economies is the*

assurance that every member of the community benefits and has enough . . . Resources are shared through a sophisticated system of distribution that takes into account the needs of every individual as well as of the whole community.

3. ***Nature Is a Source of Knowledge.*** *Indigenous communities thrive by listening carefully to the shifting patterns of nature. Natural resources are most useful and abundant when they are cultivated in harmony with the laws of nature . . . [Nature] is a model to emulate rather than a force to overcome.*
4. ***Sustainability and Resilience.*** *The evidence of Indigenous sustainability is in our thousands of years of balanced existence. It is in the health and abundance of living things on our land.*[1]

Over countless generations, through trial, experiment, and close observation, people developed sophisticated techniques for interacting with their environment that made the land more productive and their lives more secure. They used virtually everything in one way or another for food, shelter, medicine, clothing, utensils, boats, implements for fishing, hunting, and horticulture, and anything else they needed to lead comfortable lives.

Each tribe had its own social systems, attitudes, beliefs, and mythology. Indigenous people also enjoyed cultural pursuits such poetry, art, music, dance, and crafts. These were egalitarian societies in that everyone worked, everyone's voice was heard, and when times of suffering came, everyone suffered equally. Children were cared for by the entire community, as were the elderly, so there was cradle-to-grave security. They had laws but not invented laws that were based on crime and punishment. Their laws developed over time and were adopted because they effectively held the tribe together.

A commonly held misconception is that early people struggled in a pitiful existence because of fear and their incessant need to find food. Actually, tribal people led leisurely and relatively secure lives working far less than we do today. In his book *Stone Age Economics*, Marshall Sahlins refers to these Indigenous people as "the original affluent societies." They achieved this affluence by keeping their population within the land's ability to support them, and keeping their material desires to a minimum. They took only what they needed and no more. By requiring little, they did not have to work hard to live in a state of sustained prosperity. This is sometimes referred to as an "economy of abundance" because people almost always had everything they needed.

The cultures of Indigenous people arose spontaneously and developed during the many thousands of generations people have lived on the earth. It was more of an evolutionary process than a succession of learning. The things that were adopted by a particular tribe were those that gave the tribe their best chance for survival. The Indigenous people who lived ten thousand years ago, and those who still survive today, were the recipients of the sum total of all the knowledge, skills, technology, and cultural wisdom that had accumulated from earliest times and was passed down to them in an unbroken chain.

In this context it is interesting to consider Mr. Fukuoka's view of what constitutes true culture:

> *. . . wisdom acquired over time in the course of daily life . . . is simply natural wisdom as apprehended by primitive people, and should be sanctioned as wisdom bestowed by heaven . . .*
>
> *Culture is usually thought of as something created, maintained, and developed by humanity's efforts alone. But culture always originates in the partnership of man and nature. When the union of human society and nature is realized, culture takes shape of itself. Culture has always been closely connected with daily life, and so has been passed on to future generations, and has been preserved up to the present time . . . True culture is born within nature, and is simple, humble and pure.*[2]

Sense of Place

Tribal people saw the world as whole and interdependent, making no distinctions between themselves and other forms of life. They were intimate not only with the plants and animals that shared the same territory, but with natural features such as boulders, mountains, rivers, and the soil. Everything was alive and sacred. Each creature had consciousness and its own thoughts, feelings, and spirit force. When Jaime de Angulo, a physician who lived among the Pit River people for a time in the 1920s, asked one of them, a fellow named Bill, how to say "animal" in his language, Bill replied, "'Well . . . I guess I would say something like *teequaade-wade toolol aakaadzi* which means 'world-over, all living.' . . . I guess that means animals, Doc.' Jaime then said, 'I don't see how, Bill. That means people, also. People are living, aren't they?' Bill said, 'Sure they are! That's what I am telling you. Everything is living, even the rocks, even the bench you are sitting on . . . Everything is alive. That's what we Indians believe. White people think everything is dead.'"[3]

People could communicate with and learn directly from other creatures. M. Kat Anderson, in her brilliant book *Tending the Wild*, about how the Indians of California viewed the world and interacted with their environment, writes, ". . . all non-human creatures are 'kin' or 'relatives' . . . the plants, the animals, the rocks, and the water—are people. As 'people,' plants and animals possessed intelligence, which meant that they could serve in the role of teachers and help humans in countless ways—relaying messages, forecasting the weather, teaching what is good to eat and what will cure an ailment."[4]

Indigenous people considered themselves to be as closely related to plants and animals as they were to their own family. A Pomo elder, Lucy Smith, talked about how she learned how to care for her relatives from her mother when she was young:

> *[She said] we had many relatives and . . . we all had to live together; so we'd better learn how to get along with each other. She said it wasn't too hard to do. It was just like taking care of your younger brother or sister. You got to know them, find out what they liked and what made them cry, so you'd know what to do. If you took good care of them you didn't have to work as hard. When that baby gets to be a man or woman they're going to help you out. You know, I thought she was talking about us Indians and how we are supposed to get along. I found out later by my older*

sister that mother wasn't just talking about Indians, but the plants, animals, birds—everything on this earth. They are our relatives and we better know how to act around them or they'll get after us.[5]

The Indians also felt a profound connection to the place where they lived. They knew every feature of the landscape and most believed that their people had *always* lived there. Their creation myths reach back to the very beginning. Often their myths suggest that creation occurred within their own tribal territory.

They had specific names for all of the plants and knew how they could be used to provide the things they needed. Many of them were used for food, others provided fiber for clothing, cordage, and nets. Reeds and grassy plants were used for making boats, baskets, and other household items. Still others were used as medicine or for ceremonial and artistic purposes. Many plants had multiple uses. The Indians not only differentiated the various species of plants, but also identified individuals within the species for their particularly desirable attributes. Certain plants of the same family that were growing under different conditions had different properties, and they knew when and how to harvest them to get exactly what they needed.

The Indigenous people also had a complex understanding of the animals of the area wherever they lived. They knew the uses and behaviors not only of the large hoofed animals, but also of the worms, insects, small mammals, fish, and fowl. They learned this in the same way they learned about the plants—by studying their behavior and how they reacted to human intervention. Over countless generations of doing this in the same location they learned how to tend the plants and animals so they would best provide what they needed while ensuring that the land's productivity was not diminished. This knowledge was passed from one generation to the next through stories, songs, ritual ceremonies, and practical instruction.

Boundaries of each tribe were carefully drawn. Stephen Powers, a nineteenth-century anthropologist, wrote about this in 1877:

The boundaries of all tribes . . . are marked with the greatest precision, being defined by certain creeks,

> *canyons, boulders, conspicuous trees, springs, etc., each of which has its own individual name. Accordingly, the squaws teach these things to their children in a kind of sing song . . . Over and over, time and again, they rehearse all these boulders, etc., describing each minutely and by name, with its surroundings. Then, when the children are old enough, they take them around . . . and so faithful has been their instruction, that [the children] generally recognize the objects from the descriptions given them previously by their mothers."*[6]

Depending on the topography, climate, and other conditions, the tribes often migrated within their territory. Sometimes these migrations were seasonal—for example, living at high elevations during the warm seasons and moving to lower elevations when it got colder, or traveling to where a particular resource was seasonally available—but this was not random wandering, as the early Europeans believed. It was a strategic schedule to gather what the tribe needed to live and to perform "maintenance" to ensure future harvests. As time went on they shaped the landscape so it dependably provided everything they needed and less and less effort was required to maintain it.

Ingenuity and restraint at the harvest were the primary means for ensuring that vital resources were not exploited beyond their ability to recover. Anderson points out that the Indians all over California observed two "overarching" rules at the harvest: (1) Do not take it all, leave some of what is gathered for other animals; (2) Do not waste what you have harvested.[7] Seeds were collected in such a way that some fell to the ground, continuing the cycle for the following year. Bulbs were dug so the small "bulblets" would remain in the soil, providing continuity of the crop. Some plants were left unharvested for several years to give them enough time to regenerate. The harvests often amounted to a thinning process that stimulated the overall vitality of the stand.

The California Indians, as well as most tribal people worldwide, lived with an attitude of courtesy and mutual respect. Anderson explains how this played out during the harvest: "The rituals that surrounded the act of harvesting, hunting, or fishing were as important as the act itself. How one approached a plant or animal—with what frame of mind and heart—was very significant. A personal connection was often made by saying a silent prayer, leaving an offering, and thanking the plant or animal for the gift of life."[8]

The Indians were aware that they needed to limit their population to a level that did not overtax the environment. To achieve that, they used methods such as extended breast-feeding and ingesting medicinal herbs that prevented conception. With other species, the population is determined almost entirely by the availability of food. The greater the food supply, the larger the population. When the food diminishes, so does the population. The density of Native populations also reflected this.

The available supply of food in a given area is primarily determined by environmental factors. As the Indians became more adept at encouraging the wild land to produce more food their population grew in proportion, but there was always a limit beyond which they could not grow. This carrying capacity was determined by the smallest amount of food available in the leanest years.

In *The Natural World of the California Indians*, Heizer and Elasser discuss the population densities of the various tribes who lived along the Klamath River in northern California and eastern Oregon. The Yurok, Hupa, and Karok lived closest to the mouth of the river where there was an abundance of rainfall, salmon, game, and food-producing plants, notably oak trees. Moving upriver, the climate became drier, the topography became high desert, the oak trees disappeared, and large game became more scarce. This is where the Pit River tribes (Achomawi and Atsugewi) and the Modoc lived. "The figures [of population density] are Yurok, 4.66 persons per square mile; Hupa, 5.20; Karok, 2.42; Achomawi, 0.70; Atsugewi, 0.30; and Modoc, 0.30. Such population densities directly reflect the productiveness of the land in terms of available food resources. The richer the land, the more people, and vice versa."[9]

Today there are more than seven billion people in the world with many areas well beyond their carrying capacity. We compensate for this by shipping food from areas that produce more than they need to the places that cannot produce enough, or people migrate from areas of scarcity to areas of abundance. Neither of these options was viable for the Indian tribes of North America. Each tribe had to be self-sufficient. Sometimes food moved from one tribal area to another through trade, but that was only on a very limited basis. While there were ways needy tribal members were taken care of within their own community, there was no expectation that one tribe would bail out another in time of crisis. Tribal boundaries were strictly enforced so migration was not possible without creating conflict. Each tribe's members had to learn to live as best they could within the physical limits of their environment.

Techniques for Increasing Abundance

Over many generations Native people learned how to manipulate the plants, animals, and habitats in their environment to increase productivity and diversity. These activities were not "management" per se, but rather a collaboration with nature. Anderson refers to this relationship as "tending the wild" since that term does not imply control. Some of these traditional techniques included periodic and regular burning, pruning, coppicing, sowing seeds, transplanting young plants, and digging. Of these, burning was the most widely used and most effective tool.

The Indians regularly carried out large and small burns for specific purposes. These low-intensity fires were created in such a way that they did not get out of hand and did not harm trees and other large vegetation. Burning released nutrients to the soil, stimulated regrowth—which provided forage for game and other wildlife—and prevented a buildup of fuel, which would otherwise result in disastrous wildfires. It also created and maintained prairies and meadows, increased the abundance of food-producing bulbs and grasses, enhanced the density and diversity of plants, reduced competition, and helped control insects and disease.

They sowed seeds and transplanted seedlings to extend their range, and pruned existing trees and shrubs to remove deadwood, thereby increasing their vitality, life span, and productivity. Others were coppiced. Coppicing is a technique where certain trees or shrubs are cut close to the ground. The new shoots that appear from the stump grow long and straight and have fewer nodes and side branches. These shoots were particularly useful for creating building material and for making baskets, which requires long, young shoots. Occasionally the Indians weeded under particularly productive trees and shrubs to reduce competition and keep them healthy.

The effect of these practices, especially burning, was to maintain large and small areas at a mid-succession stage by simulating natural disturbance. The disturbances often reduced the dominance of existing communities, creating openings for colonization by other species. While the biomass was reduced temporarily, it was more than made up for by the increased vigor of the new growth. Some areas were burned every two or three years, others every five or six, while others were burned every fifteen years or so.

The Indians also cared for the trees, shrubs, grasses, and *geophytes** according to a regular schedule of maintenance. As long as the work was done according to schedule it did not require much effort and was not considered a burden. In fact, by all accounts Native people had plenty of leisure time. The Europeans saw that they were not working all the time and concluded that the Indians were lazy. Actually, they worked only as hard as they needed to.

When the Europeans first came to North America they were struck by the beauty and abundance of the landscape. They often described it as being like a vast garden, park, or orchard. They viewed the landscape as being pristine, "just as God had created it." They did not realize that the "savages" they found living there had largely created the landscapes they were enjoying or that they were carefully maintaining it so it would remain that way. The Indians were practicing a form of farming and horticulture that was so sophisticated the Europeans did not even notice it.

Unfortunately, this period in human history, this superb demonstration of how humanity could exist for tens of thousands of years in close partnership with nature, has all but been erased from modern consciousness. When people compare life today with the way they imagine people were living before civilization they often say something like, "Well, I certainly don't want to go back to living in caves and scrounging for food all the time." For them, it is either the modern way, or a pitiful existence in caves with constant fear of starvation and being attacked by animals. That's not the way it was at all. Maybe it was something like that during the earliest years of human existence, but at least since people gained cognitive abilities

---

* *Geophytes* are plants that store their energy underground, such as bulbs, corms, tubers, and rhizomes.

about 150,000 years ago they have lived freely as part of nature with nearly complete food security.

The Ways of Native People and Natural Farming

The tribal people living in North America had the same sensibility as other Indigenous people living throughout the world, but the individual cultures that developed, and the techniques each used to provide for itself, were uniquely molded by the environment in which they lived. In California alone there were about sixty main tribes with innumerable small tribelets. The geography of California is so varied that all of them needed to develop their own methods of "tending the wild," but they shared a similar view of the world and practiced the same ethics.

Shortly after *The One-Straw Revolution* was published in English it was reviewed in *Akwesasne Notes*, an independent Native American activist journal that originated on the Mohawk reservation in upstate New York. Referring to Mr. Fukuoka, the review reads in part:

> *Although the process he advocates arises in southern Japan, and it utilizes crops appropriate to Japanese climate and culture, the philosophy and practice of the technique is amazingly close to that of Native peoples prior to the introduction of European agriculture . . . His message is timeless and speaks to the nature of human existence. He is a Natural World philosopher, a man with an enormous appreciation of the forces of Creation, and one who understands the potential (and historical) follies of the human mind . . . his message could have been spoken by a Lakota, a Seneca, or a Zuni traditionalist. That this specific message comes from Japan is a powerful indicator that Natural people have a strong common bond throughout the world.*[10]

The complete review is reprinted in appendix B.

Despite the similarities between natural farming and Indigenous ways, there are differences, as well. Mr. Fukuoka grew up in an agricultural society. He had the typical upbringing of other children in his village, which included the cultural indoctrination of its values, practices, and acceptable social conduct. The vision he had at the age of twenty-five was Mr. Fukuoka's first glimpse at the world of non-discrimination. For some unknown reason the barrier that had separated him from the world as it actually is disappeared. He understood that people can never understand nature through the intellect.

Native people were born into that understanding. It was all they knew from birth and all that their culture knew from the beginning.

Mr. Fukuoka tried to explain his idea of the interdependence of all phenomena to his friends and co-workers but without success. They were still living in the world Mr. Fukuoka had left behind. Rather than trying to continue explaining using words alone he returned to his family farm to create a tangible example of his vision and so demonstrate its potential value to the world.

No one he knew had ever tried that kind of thing before. Over the next thirty years he pressed on, one trial after another. After many years of lonely effort he finally achieved his goal: a natural farm that validated his vision. It looked and felt like no other farm in Japan. When he began, however, he had no human guide and no idea of where his journey would lead him.

Native people had a sophisticated knowledge of nature that was handed down from generation to generation. They did not have to keep reinventing everything from the very beginning.

When Mr. Fukuoka returned to his family farm he found that the fields and the orchard were in a run-down condition. The hillside land was eroded to bare subsoil, and there were few plants growing except the heavily pruned citrus trees and a few weeds. Besides working out a natural farming method he also needed to rehabilitate a damaged landscape.

Native people inherited a pristine ecosystem. They did not have to repair damage that was previously caused by others.

Most people consider *wilderness* to refer to an area that is largely uninhabited and has not been affected by human activity. This concept is unknown to Native people. In many tribal languages the word for "wilderness" does not even exist. Present-day Indians use the term to describe land that has been neglected for a long time.[11]

All creatures need to interact with the environment in order to live. We need to do at least *something*. Even plants work, in a way, by producing what they need through photosynthesis. Early in human history people survived through harvesting and hunting alone. Once cognitive abilities and reasoning developed, our interaction with the environment became more complex and people became agents for shaping the landscape. Nature responded by giving even more than it had before. The plants that thrived were those that adapted to the care people were giving them. It is as if the landscape, over time, became tailored to its human guardians. As that happened, people came to regard the place where they lived as "home." The relationship developed this way partly because of the techniques they used, but also because of their attitude. They humbly gave thanks, fit in with other creatures, acted with courtesy, and did not try to impose their will strictly for their own benefit.

First, people chose the place where they would live. Over time, through familiarity and by lovingly tending their gardens, they became intimately connected to the land, and the land embraced them. This is what Harry Roberts referred to when he said that once you truly become part of a place, "You can feel the earth loving you back." When this relationship occurs, life is joyful. When people attempt to dominate, the land does not embrace them and life becomes a struggle.

There are divergent schools of thought about wilderness in modern society. The dominant attitude is that nature exists to be conquered and exploited. At the same time there are others who want to preserve nature in its original, pristine condition. The first is a selfish, people-first attitude that sees nature only as a commodity, while the second believes that "pure" landscapes are only those that have not been defiled by human activity. Both outlooks reveal an estrangement from nature.

Mr. Fukuoka was at home in the world without any sense of self-consciousness. He did not just sit back and admire nature, but he did not exploit it, either. He realized that nature's abundance was a conditional gift that was his only as long as he used it judiciously, and it seems that Indigenous people felt the same way.

The Indians were not visitors and they were not exploiters. The way they lived and viewed their place in the world, I believe, is the purest expression of natural farming and goes right along with Mr. Fukuoka's understanding. They both proceed from a worldview that informed their actions. Both developed a way of providing for their needs that was uniquely suited to where they lived. Those practices increased the productivity of the land and protected their source of their sustenance. They saw nature as guide and teacher and invited all forms of life to live with them as one family. And they lived with a spirit of humility and grace.

Visitors to our farm in Shuzan. *Top row left to right:* Bill; Kerry, a visitor from the Bay Area; a policeman holding Taichi; the chief of police. *Bottom row:* the local mayor and the author. PHOTOGRAPHER UNKNOWN.

The author with a basket of recently harvested vegetables. PHOTO COURTESY OF BILL DEAN.

The author digging bamboo roots on Suwanose Island. PHOTOGRAPHER UNKNOWN.

Rice hanging to dry in the Shuzan Valley.

Visitors to Shuzan from Mr. Fukuoka's farm. *Left to right:* Tsune-san, the author, Hide-san, Yu-san. Photo courtesy of Bill Dean.

A typical scene near Mr. Fukuoka's farm. Some of the rice has already been harvested and is hanging to dry. The village is in the foothills above the most productive farmland.

The hut where Yu-san and the author stayed.

One of the mud-walled huts in the orchard with a weedy ground cover, a mandarin orange tree, and acacia trees above.

Mr. Fukuoka making tea inside one of the huts.

Mr. Fukuoka standing among radish and mustard blossoms.

After twenty-five years without tillage, the soil in Mr. Fukuoka's fields has regained the layered structure of natural grasslands.

Harvesting the rice using hand tools.

Mr. Fukuoka with two giant sequoia trees near French Meadows.

Harry Roberts (*right*) showing Mr. Fukuoka some of the vegetation growing near the Green Gulch Farm.

Mr. Fukuoka examining the weeds growing in the path of the Botanic Garden at UC Berkeley.

Dr. Richard Harwood explaining the experiments they were doing at the Rodale Research Farm in Kutztown, Pennsylvania.

The Fukuokas visiting the author's parents in Los Angeles. *Left to right:* Irving and Vivian Korn, Ayako and Masanobu Fukuoka, the author.

Left to right: Wes Jackson, Masanobu Fukuoka, and Bill Mollison at the 2nd International Permaculture Conference in Olympia, Washington, 1986.

A maple sugar farmer explains the process to Mr. Fukuoka at his farm in western Massachusetts.

Mr. Fukuoka examining native California plants at the Berkeley marina with San Francisco and the Golden Gate in the background, 1986.

CHAPTER 6

Traditional Japanese Agriculture

TRADITIONAL JAPANESE AGRICULTURE, which was practiced from roughly 1600 until the end of the Second World War, was one of the finest examples of sustainable food production in a large agricultural society ever devised. Farmers combined a veneration of the land with ecologically sound practices to allow them to survive for hundreds of years without running down their environment. This is in contrast to other agricultural civilizations throughout the world that turned their homes into deserts through overuse and carelessness. What made the Japanese experience so different?

Historically, Japan was unified for the first time in 1603 under the Tokugawa clan, which ruled until 1868 with total authority. The Tokugawa, or Edo, period was characterized by economic growth, rigid social order, an isolationist foreign policy, environmental protection, and development in the arts.

In the early years, however, the Japanese were faced with serious environmental challenges. The seventeenth century was one of exponential growth. About two hundred castles with associated communities were built, creating an unprecedented demand for lumber. People also needed wood for heating, cooking, making tiles, pottery, and iron products. Farmers needed green material to fertilize their fields and to feed domestic animals. This intense demand led to widespread deforestation.

With deforestation came erosion, flooding, landslides, and a decline in water quality. The soil was also overtaxed, leading to a loss of fertility and productivity.

Other cultures have made unfortunate decisions when faced with environmental challenges of this kind, causing their civilizations to collapse and even disappear. Japan took a different course by suddenly veering onto a path of conservation and sustainability, adopting a way of life based on moderation, working cooperatively, producing no waste, and recycling. The environmental damage was first halted, and then reversed. Deforestation was controlled, replaced by an ambitious reforestation program. The fertility of the soil was restored, productivity increased, streams again ran clear, and the fish returned.

The land area of Japan is about the size of the state of Montana. More than 70 percent is mountainous and only about 15 percent can be used for agriculture so almost everyone in the rural areas lived close together in the plains and the narrow mountain valleys. Because of the large population, about thirty million in 1720, every inch of productive farmland had to be used efficiently and intensively to supply basic needs. Working together was the most efficient way to achieve that goal. The Japanese believed that the strength of many people working together is greater than the results that can be achieved by the same number of people working independently.

This sense of self-sacrifice, group identification, and cooperation is reflected in a strong work ethic. For the Japanese, work within the community is an end in itself. It is a way of life, simply what one does. Life without work in the traditional village was unthinkable. Even modern Japanese thrive on work and often do not want to stop even when they reach retirement age.

The isolation of the Japanese islands also helped mold the character and social structure of the people. Physically, of course, Japan is isolated simply by being a chain of islands. In 1635 an edict was issued that made it illegal for anyone to either enter or leave Japan. During the entire Tokugawa period Japan was isolated from the rest of the world physically, politically, and culturally. Without trade or the ability to import resources from the outside, they needed to become completely self-sufficient, and they did.

Satoyama

The arrangement of the traditional village and the state of mind of its rural inhabitants during the Tokugawa period is often referred to by the term *satoyama*, or "mountain hamlet." It suggests the picturesque setting of villages nestled between the forest and terraced rice fields, with vegetable gardens, orchards, woodlands, ponds, and clear flowing streams. It also refers to the intimate, almost reverential bond people enjoyed with their surroundings. They recognized their dependence on the land and the sanctity of the interdependence of species, and in so doing experienced a kind of spiritual fulfillment. The corresponding term for villages on the coast is *satoumi*, or "village by the sea."

The landscape was completely transformed by human activity, but human control was not complete. Many gaps were intentionally preserved to accommodate other species and allow them to flourish. The natural marshes were replaced by rice paddies that were reliably re-created each year. Most of the species that originally existed in the marshy lowlands easily adapted to these similar, yet human-created conditions. Some species died out, but those that did adapt thrived in even greater numbers. The ponds and irrigation channels were a boon to fish, amphibians, aquatic plants, small mammals, and birds. A balance existed between the cycles of growth and decay, and a reserve of organic matter was carefully maintained in the soil.

The village typically consisted of forty to one hundred households located just above the floodplains. The lowlands were reserved for growing rice during the summer and barley, rye, or a cover crop in the winter. An average holding for each family was about two and a half acres, but the plots were often scattered. Vegetables and grains were farmed on the lower slopes near the village. Orchards and specialty crops such as tea and mulberry trees for silkworms were grown in the woodlands, with forests above. Narrow paths connected the fields with upland plots. There were no fences and very few grazing animals.

The small size of the villages and fields was ideal for good husbandry. It was large enough to encourage mutual assistance on large projects like planting and harvesting the rice, maintaining the irrigation works, building a house, or thatching a roof, yet small enough to carefully maintain it using human and animal labor. The villages were largely autonomous. As long as the villagers paid their taxes, a hefty 40 percent of the rice crop, and did not otherwise attract attention, they were left alone by the authorities.

Like the country as a whole each village was almost entirely self-sufficient, and within the village the members of each household produced nearly everything they needed by themselves. In addition, each household produced a certain crop in excess or provided an exceptional skill, which was then bartered. For example, one farmer might have the only persimmon trees, but he supplied the entire village in exchange for items he could not produce himself. Another farmer might have the only fields capable of growing a specialty crop like taro. Some households produced items out of hemp or straw, made textiles, repaired tools, or kept the draft animals that were used in turn by everyone in the village. The only item that was not readily available in most areas was salt.

Repairing, reusing, and recycling *everything* was so ingrained in the way of life that they did not even have a word for "recycling."[1] In his insightful book *Just Enough*, about sustainability in Tokugawa Japan, Azby Brown writes

> *Agricultural waste—what little there is, since most plants, from root to stalk, are fully utilized in some way—becomes compost and mulch. Similarly, fireplace ash is recycled into the fertilizer mix, as are worn-out woven rush and straw items. Metal (predominantly iron) is successively reworked. A broken cooking pot may be converted into several sickle blades, for instance, and broken blades beaten into straps and hooks.*
>
> *Wood has a particularly long life cycle: a broken plow frame can become an axe handle, a broken axe handle refitted into a scoop, a broken scoop added to the firewood pile (and its ashes finding their way into the fields again as fertilizer). Clothing can be endlessly reworked . . .*[2]

The Tokugawa period was one of unprecedented peace and stability. Foreign conflicts were avoided and there were no military challenges to the shogun's authority at home. The Tokugawa leaders believed their rule would continue indefinitely so they took a long-term view of resource management. Villagers presumed their land would pass to their heirs so they, too, had incentive to keep the land healthy and productive. The preoccupation with forestry during this time, a long-cycle form of cultivation, reflects this commitment to the future.

The fact that this *satoyama* system thrived for so long demonstrates that a rich diversity of plants and animals *can* successfully coexist with intensive agriculture even in a country as densely populated as Japan. Brown summarizes this way of life as follows: "More than anything else, this success was due to . . . an understanding of the functioning and inherent limits of natural systems. It encouraged humility, considered waste taboo, suggested cooperative solutions, and found meaning and satisfaction in a beautiful life in which the individual took just enough from the world and not more."[3]

Rice-Growing

All of the most productive land in Japan was used for wet-field rice production. Sloped lands on the sides of the valley were terraced and formed into paddies wherever possible. Field sizes were rather small, up to half an acre or so; the terraced fields might be just a few square yards.

Water flowed to the fields from storage ponds in the foothills, which collected nutrient-rich runoff from the surrounding mountains. The water was delivered using gravity flow through an intricate and well-engineered system of channels and canals that reached nearly all the fields in the valley. The water would cascade from one field to the next until it was finally released back into the river. Occasionally fields were above the canals so the water had to be lifted. In those cases a human-powered pump was used.

In the spring farmers worked on the paddy walls to be sure they would hold water during the growing season, and sowed the rice seeds into carefully prepared starter beds. The main fields were fertilized with compost and then plowed to a pea-soup consistency. In June, as the monsoon season approached, the seedlings were gathered into small bundles and carried to the fields. Transplanting the rice was a team effort. The villagers moved from one household to the next until everyone's fields were planted. No money was exchanged for this work. The host family simply provided tea, snacks, and meals. The work was hard but there was a feeling of exhilaration that comes with everyone working together toward a common goal.

During the growing season the rice fields came alive. Insects, dragonflies, and amphibians appeared as soon as the fields were flooded. Fish entered the paddies with the irrigation water or were stocked. They ate insects and were in turn eaten by hawks, kites, herons, and the villagers. Egrets waded in the shallow water. Ducks swam in the paddies and perused the weeds growing on the narrow banks around the fields. Azolla, a nitrogen-fixing aquatic fern, formed a film on the surface of the water. It arrived with the irrigation water or through deliberate inoculation.

Villagers hand-weeded the fields three or four times during the growing season and lightly cultivated between the rows. Women worked side by side with men during the transplanting and harvesting, but they did almost all of the mid-season maintenance work by themselves. The harvest, also a communal effort, was done using hand sickles. Farmers hung the rice on bamboo or wooden racks for a week or two to dry before threshing. Then they plowed the field and sowed a cover crop or winter grain.

Compost

Japanese farmers were able use the land so intensively without exhausting its fertility because they returned everything to the soil, usually in the form of prepared compost. Human waste was made into compost by mixing it with crop residues and fodder they gathered from the hills nearby. Human manure was considered such a valuable commodity that many farmers set up stalls along the road hoping to attract "deposits" from passing travelers.

F. H. King, an American agronomist who traveled to Japan in the early 1900s, was impressed by the Japanese farmer's high regard for compost and the sophisticated ways they prepared it. In his classic work *The Farmers of Forty Centuries*, he wrote about how they used compost on field crops: "Manure of all kinds, human and animal, is religiously saved and applied to the fields in a manner which secures an efficiency far above our own practices . . . the largest portion of this organic matter is predigested with soil or subsoil before it is applied to the fields . . . at an enormous cost of human time and labor . . ."[4]

The common people in the cities, laborers, merchants, and artisans, lived in row houses laid out in neat grids without enough space to grow their own food. The ruling class, mainly samurai, lived in modest houses or larger estates depending on their rank. They usually had space for at least a small vegetable garden and some fruit trees. Even so, the cities could produce only a fraction of the food they needed. Therefore, a steady stream of food and forest products poured into the cities from the countryside each day. At the same time, a steady stream of farmers and laborers left the cities laden with carts of human waste. Farmers paid for this resource according to the quality of the night soil, and that was determined by the richness of the householder's diet. The waste of high-ranking samurai fetched the highest prices, low-ranking samurai were next, and the contributions of the commoners were the least expensive of all.

On a trip between Nara and Osaka in 1909, King noted:

> *As soon as we entered upon the country road we found ourselves in a procession of cart men each drawing a load of six large covered receptacles of about ten gallons capacity, and filled with the city's waste. Before reaching the station we had passed fifty-two of these loads, and on our return the procession was still moving in the same direction and we passed sixty-one others, so that during at least five hours there had moved over this section of road leading into the country, away from the city, not less than ninety tons of waste; along other roadways similar loads were moving.*[5]

Public health benefited by the removal of human waste from the cities, which kept it from piling up and polluting rivers and streams. Sewage systems were unnecessary and major epidemics did not occur during the Tokugawa period. Kitchen and bathwater were removed by a complex system of channels that reached every neighborhood. The streets were meticulously clean and the water in the canals and along the shorelines remained clean enough to support healthy communities of edible shellfish. In Europe during the same period people were dumping their waste out their windows and onto the street.

Foothills, Woodlands, and Forests

Farmsteads were located just above the valley floor. A village of one hundred households would be arranged in groups of twenty or thirty dwellings scattered along the foothills on both sides of the valley. The homes were modest affairs with the main house surrounded by outbuildings for tools, storage, and drying field crops. There was also a bathhouse, and a shed for composting night soil. Fruit trees grew in and around the courtyard. The homes were constructed using joinery rather than nails. This was convenient when repairs were needed because a single post or beam could easily be removed and replaced without replacing the entire structure. The main posts were set on stones rather than buried in the ground. This helped discourage rotting and allowed the house to shift during earthquakes.

The diet was simple yet nutritious. It consisted mainly of rice, barley, and millet, with miso soup, steamed or stewed vegetables, tofu and other soy-based foods, and pickled vegetables. They ate virtually no red meat or dairy products, and only a small amount of fish. Everything was produced locally and was eaten in season or preserved for the winter months by drying, pickling, fermenting, and cellaring. Farmers rarely ate the rice they grew because it was too valuable as a cash crop and for paying taxes. Instead they substituted millet, barley, and upland rice.

There were few draft animals and almost no livestock. This was due to the topography with its scarcity of grassland, and also to the Buddhist taboo against "eating the flesh of four-footed animals." Oxen were mainly used for the fieldwork; horses were used in the mountains. These animals would be brought out when they were needed and then returned to their small paddocks, where they might be confined for months at a time. The main animals on the farm were small, mainly ducks, chickens, rabbits, and dogs.

Vegetables grew in plots near the farmhouses. Ingenious, labor-intensive techniques were used to get the highest yields possible while still maintaining the soil's fertility. The Japanese did this by making efficient use of time and space and by returning all vegetative waste to the soil. The vegetable plots were tilled, usually by hand, and then fertilized with compost. The soil was formed into hills and planted with several crops at the same time. Then the beds were covered with mulch.

Quick-growing crops like radishes and leafy greens were harvested when they were ready and then replaced with another crop while slower-growing vegetables like tomatoes and eggplants continued to grow in the center of the beds. This technique, known as succession planting, provided vegetables continuously throughout the growing season and assured that all the land was being used productively at all times.

Vegetables like cucumbers, melons, and tomatoes were staked to bamboo trellises. Trellising took a lot of time, but it left more room for other crops. Using the vertical space as well as the horizontal expanded the overall growing area. The vegetables eventually covered the entire garden with green leaves, capturing nearly all of the available sunlight. A lot has been written in English about Asian gardening because it later became the model for the worldwide organic gardening movement.

Orchards were on higher hillsides. A wide variety of orchard trees grew in Japan because of the wide range of climates. Some of the most common orchard fruits were mandarin orange, pear, Asian pear, plum, Japanese plum, apple, persimmon, and Nanking cherry. The main nut crops were chestnut, ginkgo, and walnut. Orchard trees were usually interplanted with understory perennials, shrubs, and vines.

The orchards were surrounded by woodlands consisting of deciduous and some evergreen hardwood trees, conifers, and groves of bamboo. They were also filled with wild edible and medicinal plants, which were gathered in season throughout the year. Farmers gathered grasses, leaf litter, and other green fodder and mixed it with night soil to turn it into compost. It took five to ten acres of fodder to make enough compost to fertilize one acre of rice field.

Farmers collected limbs from fallen trees, cut them to length, and hauled them out in bundles to be used for fuel. The goal was to take the firewood they needed without compromising the health of the forest. They also coppiced oak, maple, willow, and other trees, causing new shoots to sprout rapidly from the base. This provided continuous harvesting. Some of the wood was used for firewood, but they made most of it into charcoal. Wood is more efficient for heating, but charcoal is easier to transport. In the city, where almost all of the resources had to be brought in, charcoal was the main source of fuel for heating and cooking.

The overall effect of occasional cutting and regular coppicing made the woodlands more open and productive. It made room for grasses and shrubs, which provided habitat for birds, insects, and animals that would not have survived otherwise. Many trees and shrubs of no immediate value to the villagers were allowed to grow simply to provide habitat and diversity.

Forest tree maintenance was done during the spring and summer; tree harvesting occurred during the winter. Teams of woodsmen would set up a camp and stay, often for months at a time. Workers cut and squared the logs right on the site. Then they sent them down the mountain on temporary flumes and floated them downstream. According to Azby Brown,

> *The deep forest is made accessible by sluices, chutes, trestles, dams, booms, bunkhouses, and transport vessels built en masse and on a large scale, but with few exceptions they are designed to be temporary and so are rapidly dismantled when a particular patch is logged. In most cases the structures are built of the logs themselves, which are then shipped downriver as lumber, so the industrial infrastructure largely vanishes without a trace . . . nothing is done that alters the natural functioning of the watershed.*[6]

Was Traditional Japanese Agriculture a Form of Natural Farming?

Tokugawa Japan was the epitome of a totalitarian feudal society, and life for the villagers was not always easy or comfortable. Some argue that they were exploited and oppressed, and I suppose you could see it that way. Even Mr. Fukuoka referred to them as "a poor and downtrodden lot. Forever oppressed by those in power . . . ;"[7] though he went on to say that they maintained a positive attitude and led enjoyable lives.

But was traditional Japanese agriculture a form of natural farming? To answer this question we must remind ourselves that natural farming is primarily a way of seeing the world and a way of living in which both people and nature benefit. The farming itself is an expression of how that can be accomplished.

In *The One-Straw Revolution*, Mr. Fukuoka gives four principles of natural farming. They are: (1) No cultivation—that is, no plowing or turning the soil; (2) no chemical fertilizer or prepared compost; (3) no weeding by tillage or herbicides; and, (4) no dependence on chemicals. The techniques of the Tokugawa farmers technically violated the first three of these rules. They plowed, they used *a lot* of prepared compost, and they weeded the rice fields several times during the growing season by pushing a hand cultivator between the rows. They also transplanted the rice seedlings, flooded the rice fields, and completely transformed the natural landscape. So at first glance it would seem that traditional Japanese agriculture was not natural farming at all, but that would only be considering the techniques they used.

Perhaps a complementary set of principles should be added that describes the spiritual underpinnings of natural farming: (1) View the world as a unified, interconnected, and interdependent whole that is ideally arranged just as it is; (2) respect all creatures and allow them an equal opportunity to thrive; (3) protect nature's ability to replenish itself; (4) value and encourage diversity; (5) know and care for your home; (6) take only what you need and never take it all; (7) use and recycle everything, create no waste; and (8) live with a spirit of tolerance, humility, and gratitude.

By this set of criteria the traditional Japanese rural way of life was clearly a form of natural farming. Mr. Fukuoka acknowledged as much by the admiration and respect he held for Japanese farmers and for villagers all over the world, even as he demonstrated how their farming techniques could have been improved. In many ways he considered the *satoyama* way of life to be an exceptional expression of how people could get along harmoniously with nature.

They did not trouble themselves with finding the true meaning of life or solving other ponderous metaphysical questions. Here is how Mr. Fukuoka described the philosophy of traditional Japanese villagers: "Farmers preferred to live common lives, without knowledge or learning. There was no time for philosophizing. Nor was there any need. This does not mean that the farming village was without a philosophy. On the contrary, it had a very important philosophy. This was embodied in the principle that 'philosophy is unnecessary.' The farming village was above all a society of philosophers without need of philosophy."[8]

CHAPTER 7

Organic Farming and Permaculture

NATIVE PEOPLE AND THE VILLAGERS of traditional Japan created a way of life that they thought would go on forever. They respected biological limits, took only what they needed, appreciated and lived harmoniously with other forms of life, protected nature's ability to renew itself, and were grateful for what was given to them. Today their way of life is all but gone, supplanted by modern cultures that do not share their sustainable values.

Masanobu Fukuoka had nothing good to say about modern culture. He believed that all of the problems we have today are the result of people separating themselves from nature, and in so doing losing contact with the touchstone of truth. True culture originates in a partnership between people and nature and is passed down from one generation to the next. Once that link was broken, humanity, for the first time, had to set its own course—and the results have been disastrous. Instead of building great civilizations, he believed we would have been better off not doing anything at all.

Many people consider the development of agriculture to be the cause of our current social and environmental problems, but the advent of agriculture merely coincided with a more profound change: People decided that human beings were superior to other forms of life, that nature was created for us to do with as we pleased, and that it was our destiny to conquer and rule the world. In other words, people came to believe that we were above natural law. The change was not as much about agriculture as it was about power and control, but agriculture proved to be an effective tool for spreading its influence.

Native people had no trouble deciding how to live because their lives remained largely unchanged from one generation to the next. When people separated themselves from nature they cut themselves off from their traditional sources of knowledge—intuitive understanding, learning directly from other creatures, and being guided by what generations before them had done. They could no longer see the whole of Creation; instead, they had to rely on their intellect, which can only comprehend small pieces of reality at one time. This fragmented way of seeing the world eventually coalesced into science and became the standard way for people to organize experience and decide what to do.

Tribal societies were considered affluent because they were able to produce all they needed with a minimum of effort. They could do this because they desired little more than what they needed to live. Our modern society, on the other hand, has an insatiable appetite. Instead of the economics of abundance, our highly productive system is referred to as the "economics of scarcity" because no matter how much we produce, it can never be enough. Our economic system institutionalizes the need for expansion in an ultimately futile attempt to keep up. This is known "growth," or "progress," which in its application results in extracting natural resources as quickly and efficiently as possible.

Examples are everywhere, but let me mention just one from my home state of Oregon. Until about 150 years ago, salmon and steelhead migrated up the Columbia River and its tributaries in unimaginable numbers. Native people said the fish were so plentiful during the spawning season that "you couldn't see the bottom of the rivers and streams." The Indians who lived near these waterways caught some of the fish to eat right away and preserved what they would need for the winter. What they took only amounted to a minute portion of the total number. In the mid-1800s Europeans noticed how the Indians were dipping nets into the water to catch fish, and in 1879 the first "fish wheel" appeared.

A fish wheel was a water-powered device mounted on a floating dock. It operated much like a conventional waterwheel, but the paddles had baskets on them. The migrating salmon were funneled to the wheel, where they were scooped up and deposited into a large storage bin. When the bin was filled it was emptied and the fish were taken to a nearby cannery. One wheel could harvest up to fifty tons of fish in a single year. By 1906 there were seventy-five of them operating on the Columbia River. The goal was to take as many fish as possible, as quickly as possible. Today we refer to that strategy as "maximizing production." Within two generations the once plentiful runs of migrating salmon and steelhead dwindled to a tiny fraction of what they once were. One can only imagine what the Native people thought when they saw those fish wheels in operation.*

In a 2008 article on the history of Columbia River fisheries written for the Northwest Power and Conservation Council, John Harrison said, "Viewed in hindsight, over fishing appears reckless and greedy, but in the context of the time this was not the case . . . in that era it was not uncommon to believe that the supply of all natural resources—fish, trees, water, and land for agriculture—essentially was limitless." True, but even today when we are aware of environmental limits and the consequences of overharvesting, the take-it-all-as-fast-as-possible mentality is still with us, and will continue to be until society fundamentally changes the way it sees itself in relation to the natural world.

Not all of the Europeans who came to North America were intent on plunder, although that was, and still is, the driving ethos of American culture. Wendell Berry describes the two major tendencies in our society as the exploiter and the nurturer:

> *I conceive a strip-miner to be a model exploiter, and as a model of nurture I take the old-fashioned idea or ideal of a farmer. The exploiter is a specialist, an expert; the nurturer is not. The standard of the exploiter is efficiency; the standard of the nurturer is care. The exploiter's goal is money, profit; the nurturer's goal is health—his land's health, his own, his family's, his community's, his country's. Whereas the exploiter asks of a piece of land only how much and how quickly it can be made to produce, the nurturer asks*

---

* Fish wheels were banned in Oregon in 1928 and in Washington in 1935.

> *a question that is much more complex and difficult: What is its carrying capacity? . . . The exploiter wishes to earn as much as possible by as little work as possible; the nurturer expects, certainly, to have a decent living from his work, but his characteristic wish is to work* as well *as possible.*[1]

Within a few hundred years of the Declaration of Independence nearly all of the nation's arable land had been settled by people who decided to stay in one place and set down roots. Most of these settlers did not aspire to take all they could, but only what they needed to live. They worked to protect the land as best they knew how so it would be passed on to their heirs in good condition.

While most of these small-scale diversified farmers have been replaced by agribusiness mega-farms, there are a growing number of farmers scattered throughout North America who, against all odds, are managing to live on the land according to traditional rural values, enjoying all the benefits and responsibilities of country life. Others are doing so even though they live in the city or the suburbs and have only small patches in which to garden. Many of these people are part of the growing organic farming and permaculture movements.

Organic Agriculture

Although scientific agriculture has become the dominant form of agriculture worldwide, not everyone headed down the road to industrial agriculture. Some had a different, more holistic idea about the relationships among soil, plants, and human society. They believed that the way we grew our food was crucial to maintaining health. While different people worked on different aspects of this issue, all acknowledged the interconnected nature of living things and the importance of maintaining a sufficiently high level of organic matter, or humus, in the soil. They believed that production should not be the singular goal of agriculture because that separated cropping from the complete cycle of life, or Wheel of Life, which consists of birth, growth, maturity, death, decay, and rebirth.

If only agriculture would adopt techniques that protected the soil, they believed, the land would be able to support a thriving human population indefinitely. "The remedy is to look at the whole field covered by crop production, animal husbandry, food, nutrition, and health as one related subject and then to realize the great principle that the birthright of every crop, every animal, and every human being is health."[2]

F. H. King, Sir Albert Howard, and J. I. Rodale

After a long and distinguished career as a soil physicist at the University of Wisconsin, Franklin Hiram King retired in 1908. Not one to sit idle, he used money from his life insurance policy to fund a nine-month trip to China, Korea, and Japan. He wanted to examine firsthand how the agricultures of Asia had successfully maintained such dense populations for thousands of years. These travels are recorded in King's best-known work, *The Farmers of Forty Centuries: Permanent Agriculture in China, Korea and Japan.*

He interacted with his hosts as equals, as people who could teach him how to the solve problems of soil depletion and erosion that farmers in the United States were experiencing but had never anticipated. He repeatedly remarked that Asian techniques for such things as water management, cover-cropping, and efficient use of time, space, labor, and materials were superior to those of the West. He wrote:

> *In selecting rice as their staple crop; in developing and maintaining their systems of combined irrigation and drainage . . . in their systems of multiple cropping; in their extensive and persistent use of legumes; in their rotations for green manure to maintain the humus of the soils and for composting; and in the almost religious fidelity with which they have returned to their fields every form of waste which can replace plant food removed by the crops, these nations have demonstrated a grasp of essentials and of fundamental principles that may well cause western nations to pause and reflect.*[3]

King returned to the United States and immediately began work on the manuscript, but he died suddenly on August 4, 1911, before he had a chance to write the final chapter. Typesetting began the next day anyway under the able supervision of his wife, Carrie Baker King. It was self-published by Mrs. King in 1911. The book languished until it was rediscovered in the 1930s by Sir Albert Howard, Jerome Irving Rodale, Lady Eve Balfour, and others who would join to initiate the worldwide organic farming movement. Rodale republished the book in 1949 and it has remained in print continuously ever since.

The person most responsible for articulating the principles of the organic farming movement was Sir Albert Howard (1873–1947). Throughout his life Howard published many books and articles. His best known are *An Agricultural Testament* (1940) and *The Soil and Health* (1947). They were written with both general readers and scientists in mind. Howard grew up in the English countryside and was trained as a mycologist at Cambridge University, in London. In 1905, after spending a few years working in the West Indies and a few years teaching agricultural science in England, he traveled to India where he would spend the next twenty-six years of his life directing agricultural research.

Howard's first appointment was to The Research Institute at Pusa, near Calcutta. Since he was not familiar with the farming in India, he spent most of his time there learning from local farmers, whom he referred to as his "professors." He watched them produce healthy crops of wheat, chickpeas, and tobacco without using chemical fertilizer or insecticides. Howard also noticed that the draft oxen used at the institute did not suffer from the contagious diseases that plagued animals on the neighboring farms even though they were in such close proximity that they rubbed noses across the fence lines.

"From these observations on plants and animals, Sir Albert was led to the conclusion that the secret of health and disease lay in the soil. The soil must be fertile to produce healthy plants and fertility meant a high percentage of humus. Humus was the key to the whole problem, not only of yields but of health and disease. From healthy plants grown on humus-rich soil, animals would feed and be healthy."[4] To replace the humus removed from the soil, Howard turned to composting, crediting the Chinese with the idea. In a memorial article for her husband, Louise Howard wrote, "On this crucial question of returning wastes to the soil, he always acknowledged his debt to the great American missionary, F. H. King, whose famous book, *The Farmers of Forty Centuries* ... was to him a kind of bible."[5]

The Chinese system of agriculture described by King and Howard became the model for the worldwide organic farming movement as popularized by J. I. Rodale through *Organic Gardening* magazine and countless other Rodale publications. Two linked features characterize this system—plowing, and lots of work. The decomposition of organic matter in the soil occurs through the process of oxidation, similar to digestion in the human body. The rate of this slow, steady burn is regulated by the amount of oxygen in the soil. In a natural soil the rate of decomposition matches the amount of plant material the soil produces along with the droppings and decaying bodies of animals and microorganisms. When the soil is plowed, the amount of oxygen is increased so the rate of decomposition increases. To maintain the fertility of the soil, new organic matter must be added on a regular basis. That's where the work comes in, and the need for all that compost.

When human beings first learned to plow, they gained access to the vast reserve of solar energy that had been stored in the organic matter of the soil, but access to this energy came at a very high price in the form of labor, erosion, and other environmental consequences. Albert Howard thought the trade-off was worth it because it allowed civilization to flourish. "[Through cultivation of the soil] man has laid his hand on the great Wheel and for a moment has stopped or deflected its turning. To put it another way, he has for his own use withdrawn from the soil the products of its fertility. That man is entitled to put his hand on the Wheel has never been doubted, except by such sects ... who argued themselves into a state of declaring it a sin to wound the earth with spades or tools."[6] He believed that it was perfectly all right to plow the soil, entirely remaking nature in the process, as long as people also put in the hard work to maintain the soil's fertility. "All the great agricultural systems which have survived have made it their business never to deplete the earth of its fertility without at the same time beginning the process of restoration. This becomes a veritable preoccupation."[7]

Mr. Fukuoka and the Indigenous people of the world did not think it was a good idea, morally or otherwise, "to withdraw for his own use the products of the soil's fertility" by plowing. They were content to coexist with the land in a more gentle way. They also did not want to make replacing the earth's fertility "a veritable preoccupation."

This dichotomy between nurturing the land for the benefit of all species and using it strictly for the advancement of human civilization is summed up in an illuminating paragraph from Howard's *Soil and Health*:

> *What is agriculture? It is undoubtedly the oldest of the great arts; its beginnings are lost in the mists of man's earliest days. Moreover, it is the foundation of settled life and therefore of all true civilization, for until man had learnt to add the cultivation of plants to his knowledge of hunting and fishing, he could not emerge from his savage existence. This is no mere surmise: observation of surviving primitive tribes, still in the hunting and fishing stage . . . show them unable to progress because they have not mastered and developed the principle of cultivation of the soil.*[8]

In this passage, Howard reveals the smug attitudes of his culture. He maintains that until plowed-field agriculture came along, the basis of all true civilization, people led a "savage existence." In a later passage he referred to the way Indigenous people obtained their food as "nothing more than a harvesting process."[9] Without plowing the soil these primitive people were "unable to progress," and the only way they could improve themselves was to change their ways to be like his. Of course that would entail remaking their entire way of life and violating their own ethics. Perhaps these people had no interest in joining the "march to progress," did not believe that dominating nature was a good idea, and felt it was not "man's destiny" to do so. The irony is that even though things have not gone at all well for human society or the environment over the past ten thousand years, the air of superiority persists.

In 1931 Howard retired from government service and returned to England. *An Agricultural Testament* was published in 1943. His ideas were immediately attacked by the agricultural establishment, which viewed them as exaggerations and oversimplifications. He was marginalized as an extremist largely because of his lack of scientific proof and his hard-line stand against the use of synthetic chemicals of any kind. Where were the comparison plots? Where was the data? Science demanded that he use their empirical criteria while Howard's understanding was based largely on whole systems analysis, intuition, and a lifetime of experience.

Howard found a willing comrade-in-arms in Jerome Rodale. J. I. Rodale, as he is best known, was himself a fascinating person. He was an author, editor, publisher, playwright, and avid gardener. The son of a grocer, Rodale grew up on Manhattan's Lower East Side. He tried various jobs and business ventures, eventually moving with his wife, Anna, to Emmaus, Pennsylvania, in 1940. At the time, he was partners with his brother in an electrical manufacturing business. He read Albert Howard's work and was inspired to buy a sixty-three-acre farm to put Howard's then-unconventional theories into practice.

In 1942 Rodale began publishing *Organic Gardening and Farming* magazine* to help promote his belief that simple, unprocessed, organically grown food, produced in healthy soil and a clean environment, was infinitely better than the alternative. He mainly concentrated on topics that would interest backyard gardeners, such as composting, food storage, food preparation, and the care and conditions the various fruits and vegetables needed to thrive. The magazine attracted a small cadre of like-thinking gardeners and small-scale farmers but it continued to struggle for many years. Rodale's *Prevention* magazine, which dealt with the issues of health, nutrition, and fitness, debuted in 1950.

In the 1960s Rodale Press started publishing books about self-reliant living, broadening its appeal to members of the growing environmental and back-to-the-land movements. These publications provided a primary source of information and inspiration for those who were turning away from the cities to take up new lives in the countryside. Rodale suddenly found himself in the forefront of what had become a sizable movement.

J. I. Rodale died unexpectedly in 1971, but the company was ably taken over by his son, Robert. Under his leadership, Rodale Press finally came into profitability as sales soared. He broadened his father's message, redefining it as an "organic lifestyle," and put more of an emphasis on fitness, which was a personal passion of his. He also promoted the

---

* The name was later changed to *Organic Gardening*.

concept of "regenerative agriculture," which was a forerunner of today's sustainable agriculture movement. At the time Rodale Press published *The One-Straw Revolution* in 1978, the circulation of *Organic Gardening* was well over one million. Even today *Prevention* magazine still has a circulation of almost three million.

No-Tillage Organic Agriculture

Not all organic farmers believed that plowing the soil was a good idea. Edward H. Faulkner (1886–1964) was one of them. His book, provocatively titled *Plowman's Folly*, produced a storm of controversy when it was published in 1943 because it challenged the commonsense agricultural wisdom of using the moldboard plow.

Faulkner grew up in rural Kentucky, was trained in agriculture at Cumberland College (then Williamsburg Baptist College), and worked for many years as a county extension agent in Kentucky and Ohio. His unorthodox views led to his dismissal from government service. He moved to Elyria, Ohio, where he carried out experiments to test the efficacy of his theory that organic fertilizer is best added from the top of the soil rather than burying it with a moldboard plow.

Faulkner's main thesis was that plowing was unnatural and destructive. It led, he said, to a host of problems that were not present in forests or unplowed meadows. In a passage curiously reminiscent of Mr. Fukuoka's ideas, Faulkner wrote:

> *From one point of view, we have been creating our own soil problems merely for the doubtful pleasure of solving them. Had we not originally gone contrary to the laws of nature by plowing the land, we would have avoided the problems as well as the expensive and time-consuming efforts to solve them . . . we would also have missed all of the erosion, the sour soils, the mounting floods, the lowering water table, the vanishing wild life, the compact and impervious soil surfaces . . ."*[10]

He had difficulty finding a publisher for *Plowman's Folly*, but finally the University of Oklahoma Press agreed to do it. The book came out on July 5, 1943, and became a bombshell almost immediately. In a little more than a year it went through eight printings and sold more than 250,000 copies. Faulkner had struck a nerve and overnight this "soil and crop investigator in private employment" (from the dust jacket), who was actually selling insurance when the book came out, found himself in the spotlight and the subject of intense controversy.

The media picked up the story, treating it as if it were David versus the Goliath of mainstream agriculture. *Time* magazine called it the "hottest farming debate since the tractor first challenged the horse." In his review of *Plowman's Folly* for *The Nation*, Russell Lord, a Faulkner supporter, wrote, "No book . . . in the last thirty years of agriculture has aroused such a furor; and that delights me."[11]

Faulkner did not get much support from rural farmers, however. They considered his ideas as going against traditional ways and common sense. Some scientists did not want to dignify Faulkner by responding, but big-time agriculture certainly did, trotting out one expert after another to confront the heretic. Articles in popular magazines and professional journals strongly refuted his claims as being unscientific, amateurish, and, again, going against common sense. The responses took on a frenzied intensity that went far beyond the actual threat Faulkner represented, reminiscent of the treatment Albert Howard received from agricultural scientists and what Mr. Fukuoka experienced in Japan.

While some well-known soil conservationists like Hugh Bennet praised Faulkner's ideas, he found his greatest support among ordinary citizens. The tragedy of the Dust Bowl was still fresh in people's minds and they blamed the government's agricultural scientists for allowing it to happen. America was in the middle of World War II and victory gardens had, of necessity, turned the country into a nation of gardeners, with whom Faulkner's message resonated. But eventually the forces of orthodoxy prevailed and Faulkner's star faded as rapidly as it had risen.

The organic no-till advocate with perhaps the broadest vision was J. Russell Smith (1874–1966). His best-known work, *Tree Crops: A Permanent Agriculture*, was first published in 1929. A revised edition came out in 1953. Smith was, above all, a soil conservationist. He saw the devastation caused by plowing, especially on sloped land, and it made his heart bleed. His solution was first to stop plowing, because that was what was causing the problem in the first place, and then to plant trees—lots of them, and ones that produced food or fodder.

Europeans brought their plowed-field techniques with them from the old country to grow crops of wheat, rye, barley, and oats. At first they grew these grasses in the valley bottoms, which were relatively flat, but big problems arose when they added New World row crops such as cotton, corn, potatoes, and tobacco and expanded their fields to the sloping hillsides. Row crops left the soil exposed for the entire season, so when thunderstorms came, raindrops pounded the surface, dislodging the soil particles and providing the transport mechanism to carry them away. The steeper the land, the worse the problem became.

Smith traveled extensively during his life visiting examples of both destruction and successful adaptation. In China he saw barren hillsides cut by deep gullies where farming villages had once thrived. Farmers of past generations cleared the forests and then plowed the sloping land, ruining the hillsides and the floodplains as well. "Hence, the whole valley, once good farmland, had become a desert of sand and gravel, alternately wet and dry, always fruitless."[12]

In Corsica, Smith found a completely different landscape—hillsides covered with chestnut trees and fruit orchards interspersed by villages and stone houses. "These grafted chestnut orchards produced an annual crop of food for men, horses, cows, pigs, sheep, and goats, and a by-crop of wood . . . The mountainside was uneroded, intact, and capable of continuing indefinitely its support for the generations of men."[13] For flatter land he suggested a "two-story" system with trees above and pasture or cultivated crops below, a system we refer to today as agroforestry. He called it Permanent Agriculture.

His overall vision was even broader. "I see a million hills green with crop-yielding trees and a million neat farm homes snuggled in the hills . . . The unplowed lands are in permanent pasture, partly shaded by cropping trees—honey locust, mulberry, persimmon, Chinese chestnut, grafted black walnut, grafted heartnut, grafted hickory, grafted oak, and other harvest-yielding trees."[14] To achieve this kind of agriculture Smith knew that the population needed to be redistributed so the land could be tended with the care and knowledge that only one who permanently lived there could provide. He believed that a permanent agriculture could be applied anywhere, including in small yards.

Smith's vision of what society could be like is reminiscent of Mr. Fukuoka's utopian vision:

> *In my opinion, if 100% of the people were farming it would be ideal. There is just a quarter-acre of arable land for each person in Japan. If each single person were given one quarter-acre, that is 1¼ acres to a family of five, that would be more than enough land to support the family for the whole year. If natural farming were practiced, a farmer would also have plenty of time for leisure and social activities within the village community. I think this is the most direct path toward making this country a happy pleasant land."*[15]

One unusual no-till organic method for small-scale gardening was created by a fascinating woman named Ruth Stout (1884–1980). She wrote about her technique in several books, including the delightfully titled *How to Have a Green Thumb Without an Aching Back* (1955), *The Ruth Stout No-Work Garden Book* (1973), and *I've Always Done It My Way* (1975), as well as in a column she wrote for *Organic Gardening* from 1953 to 1971. The sheer simplicity of her "no-work" method and her witty, no-nonsense personality made her extremely popular among backyard gardeners and members of the back-to-the-land community in the 1960s and 1970s.

Stout, who was affectionately referred to as the Mulch Queen and the No-Dig Duchess, grew up in Topeka, Kansas, and moved to New York City when she was eighteen years old. She married Alfred Rossiter in 1934, and on a whim they decided to move to a rural property named Poverty Hollow on the outskirts of Redding, Connecticut. She loved living in the country and began gardening right away.

After fifteen years of "gardening like everybody else" she had an experience that changed the way she gardened for the next thirty-five years. When the person she had hired to plow her garden in the spring did not show up as scheduled, she felt helpless because she could not plant her vegetable garden on time. She decided to do it anyway, but on the surface of the unplowed ground. Then she covered the plot with mulch. She said she decided go ahead and do that after consulting with an asparagus plant.

Stout's method is all about mulch—lots of it. She preferred hay although she said that you could use straw, leaves, kitchen scraps, pine needles, weeds—any vegetable matter that will rot. The bed should be prepared with a layer of mulch at least eight inches high. When the mulch settles with the rain and decomposition, add more mulch. In fact, "add more mulch" was her recommendation for just about any circumstance a gardener was likely to encounter. The mulch decomposes into nutrients for the plants, keeps the soil cool during the summer, controls most weeds, and eliminates the need to water. She pulled back the mulch in the spring to allow the soil to warm up and dry out, then rolled it back. For difficult weeds, Stout sometimes used cardboard or a thick layer of newspapers under the mulch. This did not eradicate them completely, but it set them back so severely that gardening was manageable.

Like Mr. Fukuoka's techniques, Stout's uncomplicated method can be characterized nearly as well by what she did *not* do as by what she did. She did not need to plow, spade, grow a cover crop, weed, water, or spray. There was also no need for the hard work of making compost. This gave her more free time, which she enjoyed, and also allowed her to continue gardening until well into her nineties.

Stout's no-till mulch method has been compared to Mr. Fukuoka's way of gardening, and certainly there are similarities. But Stout's method is only practical in relatively small plots, while Mr. Fukuoka's approach is better suited to larger areas. Also, Mr. Fukuoka used a lighter mulch and did not rely on it entirely for weed control. But in their emphasis on no-tillage, simplicity, and the satisfaction that comes from not doing unnecessary work, the two techniques could at least be considered kissing cousins.

When I consider these two people who were the products of entirely different cultures, I cannot help but notice the things they had in common. I think they would have really enjoyed meeting each other. Stout referred to herself as "a 100% non-conformist." She was born a Quaker but later rejected religion, had no interest in politics, patriotism, or people who put on airs. She hated to leave home and never learned to drive. She taught herself Russian so she could read Dostoyevsky in the original. And she was eminently practical and self-reliant. But what brings Ruth Stout and Mr. Fukuoka together more than anything in my mind is their humanitarianism. They could have been perfectly happy doing what they did privately at their own homes, but instead they chose to share their experiences because they thought they might improve the lives of other people.

Stout addressed this with her characteristic wit:

> *At the age of eighty-seven I grow vegetables for two people the year-round, doing all the work myself and freezing the surplus. I tend several flower beds, write a column every week, answer an awful lot of mail, do the housework and cooking, —and never do any of these things after 11 o'clock in the morning! But that is my one real success, because I have had over three thousand people from every state and from Canada come to see my easy method and I have received thousands of letters, thanking me for making gardening easy for them. Or possible. And I feel that if you really help people you've done something worthwhile.*[16]

Comparing Scientific and Organic Agriculture

Organic gardening and farming originated partly in reaction to damage caused by scientific agriculture, and partly from the innate nurturing instincts of a relatively large but subdominant segment of society. Organic agriculture offers a more benign approach to producing food, but retains the fundamental outlook of modern agricultural in that it, too, begins by asking "How can I get nature to best produce for me?" The scientific farmer believes that using chemicals is the most efficient way, while the organic farmer believes that using organic material is best.

There are several forms of organic agriculture. By its narrowest definition, it is simply farming without using synthetic chemicals. This commonly held perception is overly simplistic, however. The vision of organic farming as put forth by Sir Albert Howard, J. I. Rodale, Robert Rodale, and others is much broader and more complex because it is based on the recognition of the interconnectedness of life, and the crucial importance of the health of the soil to the health of crops, livestock, individuals, and society as a whole.

If the narrow definition of organic agriculture is used, there is nothing wrong with creating enormous farms that use large-scale machinery, raise crops in monoculture, and spray the fields with "organic" amendments and pesticides. There are many farms like this today that qualify for organic certification, but they do not produce vibrant crops or nutritious food, and should be considered as nothing more than another form of industrial agriculture. Some large-scale organic farms, like the Lundberg Family Farms, are conscientiously doing their best to maintain the health of the soil and grow nutritious crops despite their size, but they are the exception.

Many people use organic farming techniques to raise food for themselves in their backyards and on small-scale farms. They are motivated by a desire to produce wholesome food for themselves, their family, and their friends, and perhaps sell at local shops, farmers markets, and through community-supported agriculture (CSA) programs that connect farmers directly with consumers. If more people grew their own food organically in urban and suburban areas it would go a long way toward establishing a healthier and more sustainable society.

Organic and scientific farming exist in a push–pull relationship. Ultimately, they are two sides of the same coin acting out their respective tendencies of exploitation and nurture. I have already pointed out that even Sir Albert Howard, who provided the broadest, most profound vision for the organic farming movement, believed that humanity had every right to remake the landscape entirely for human benefit. According to Howard it is perfectly acceptable for people to "put their hand on the Wheel" and divert soil fertility for human use if they are careful to return everything they have taken back to the soil. More often than not, however, the "return what was taken" side of the equation is conveniently ignored. A few places where an agriculture that is based on a healthy respect for the Wheel of Life *has* worked is on many Amish farms, where the community lives according to a strict code of ethics; in Tokugawa Japan, where it was built upon a spirit of reverence and respect for nature; and in isolated pockets around the world where villagers rely mainly on tree crops.

Permaculture

Permaculture is a design system created in the 1970s by Australians Bill Mollison and David Holmgren. It is based on carefully observing and then replicating natural ecosystems to create whole, self-sustaining agricultural landscapes and human communities. Since it emphasizes interrelationships rather than individual elements it is often referred to as a form of ecological agriculture. The concept goes beyond just agriculture because it integrates food production with such things as social and economic structures, city planning, access to land, cooperative banking, and social justice. The word *permaculture* is an amalgamation of *permanent agriculture* and *permanent culture*.

Permaculture systems involve a wide diversity of plant and animal species, emphasizing tree crops and other perennial plants that have many beneficial uses, or functions. The elements, which include such things as plants, animals, buildings, composting areas, and ponds, are arranged so they interconnect in as many ways as possible. This gives the system resilience and adaptability. The design is created to take full advantage of the particular location, so no two designs are alike.

One of permaculture's signature concepts is the "edible forest garden." In this system the tiered structure of natural woodlands is imitated, but with a higher proportion of edible plants and species that are useful in many ways. The plants are often arranged in combinations called guilds, which are groupings of herbaceous plants, shrubs, and trees known to get along particularly well together.

When creating the initial design the designer must look ahead, imagining what the forest garden will be like when it reaches maturity in twenty, thirty, or perhaps fifty years. Instead of waiting for the plants to go through each stage of succession one after the other, the plants of all stages are installed at the same time, speeding up the process. While the trees are still young more sunlight reaches the ground, so annuals predominate. As the trees grow larger and the area becomes shadier, the annuals give way to food produced by perennial plants, bulbs, fruit and nut trees, and berries.

The fully realized edible forest garden presupposes a rural setting, but permaculture can also be practiced in smaller urban and suburban yards. You simply take the particular conditions into account and apply the principles as best you can. There are no hard-and-fast rules. As long as you do not violate the ethics and principles of permaculture you are free to do anything you like. A permaculture farm or yard is often distinguished by its eclectic and whimsical character. Creativity is not only tolerated, it is encouraged. One of permaculture's many adages is, "The possibilities are limited only by the imagination of the designer."

Here is how the design process works. Designers carefully study the area where they will be working, preferably for at least a year so it can be experienced in all seasons. They study the wild land nearby, observing the patterns and relationships of the plants and animals that live there. Then they observe the plants and animals that live on the actual site, examine the condition of the soil and perhaps get a soil survey, check the water situation, the overall climate, and the microclimates.

They interview neighbors who tell them about the history of the area, familiarize themselves with the local codes, make a list of plants for various uses that are appropriate to the area, measure the angle of the sun and find its extreme north–south positions at the solstices, and so forth. Then they make a list of each element in the design, and analyze what each needs and what it will provide to the other elements. Finally, they sit down at a desk and create a design using a base map with overlays, along with a complete write-up. Then the design is installed, sometimes in phases, sometimes all at once.*

Around 1980 Bill Mollison came to the United States for the first time to explain and promote permaculture. At the time, people were just beginning to recognize that widespread environmental damage was occurring. Environmentalists and researchers were searching for better ways of doing things, but everyone was focused on their own field of study and did not have any particular interest in what others were doing. That kind of specialization is not possible with permaculture because it emphasizes the interconnection of all phases of living. Everyone needed to know at least the fundamentals of sustainable food

---

* There is much more to permaculture than I can cover in this short description. A number of good books have been written on the subject. If you would like to find out more, Toby Hemenway's *Gaia's Garden: A Guide to Home-Scale Permaculture* would be a good place to begin.

production, raising animals, soil and water management, appropriate technology, and natural building, as well as the social, economic, and political aspects of life.

Mr. Fukuoka came to the attention of permaculture for several reasons. In his first book, *Permaculture One*, Mollison did not even mention growing grain. He was committed to a no-tillage approach, but he did not know how to grow grain without turning the soil. Nobody did. Then he read *The One-Straw Revolution* in which Mr. Fukuoka described how he managed to do it successfully on his farm in Japan. His method was described in *Permaculture Two* and has been a fixture in the design course curriculum ever since. Also, although Mr. Fukuoka used a completely different approach, his orchard was often used as an example of what a mature edible forest garden should look like.

Another reason Mr. Fukuoka's approach is often included in course teachings is that it adds a spiritual dimension some feel is lacking in permaculture. People often combine permaculture with other disciplines or philosophies to create hybrids that personally resonate with them. Some introductory design courses combine the basic seventy-two-hour curriculum with yoga, meditation, survival skills, or biodynamics.* Biodynamics is particularly well represented; many farms refer to themselves as permaculture/biodynamic farms. Several of them are among the most inspiring I have ever visited.

When permaculture first came to the United States a number of people were skeptical. Some doubted the wisdom of throwing together such a diversity of native and exotic species, perhaps for the first time. Others wondered how planting aggressive species like acacia, Russian olive, bamboo, white clover, and honeysuckle could be compatible with the idea of minimal maintenance. There was also the question of whether or not a perennial agriculture based on tree crops could

* Biodynamics is a way of farming that grew from the teachings of Rudolph Steiner (1861–1925), an Austrian philosopher and spiritualist. It uses many standard organic practices such as making compost and cover-cropping, but adds its own techniques such as aligning farm activities with the movement of heavenly bodies and applying specially made "preparations" to the soil. Biodynamics seeks to restore and harmonize the vital life forces of the farm by working in cooperation with "subtle cosmic influences."

produce enough food to adequately replace the early-succession (plowed-field) agriculture that already existed. There were examples of permaculture in Australia, but none in the United States. Trials and practical demonstrations were needed.

The Pacific Northwest was ideally suited to answer these questions. A group of small-scale organic farmers and local-food advocates had already formed a grassroots, regional organization called Tilth. They worked to encourage more opportunities for local growers by creating farmers markets, land-sharing cooperatives, and community bartering. They also published a newsletter that was distributed in Oregon, Washington, Idaho, western Montana, and parts of British Columbia. Permaculture was immediately embraced by the Tilth group. They promoted it in their publication and hosted two workshops, one in Idaho and the other near Portland, which were well attended. In 1982 Tilth published *The Future Is Abundant*, a resource guide for those who sought to apply permaculture and natural farming in the Pacific Northwest.

It is hard to imagine better growing conditions than the Maritime Pacific Northwest. The mild climate, abundant rainfall, and deep rich soils encourage plants to grow quickly and lavishly, much more so than in arid areas. The Tilth people got to work right away and soon demonstrations of regionally adapted permaculture, both urban and rural, started popping up all over. Mollison took notice and came to the area a number of times. He taught introductory courses, visited farms, gave public talks and radio interviews, and generally encouraged the efforts that were being made. In 1986 the Second International Permaculture Conference was held in Olympia, Washington.

A lot has changed in the Pacific Northwest over the past forty years. When that ragtag Tilth group first formed in 1974 there were two farmers markets in the state of Washington. Now there are hundreds. The uncertainties surrounding permaculture when it first arrived have largely been resolved. Today thousands of working permaculture farms and backyard gardens exist throughout the region, demonstrating that a perennial agriculture *can* be highly productive. Permaculture has also been shown to be an effective tool for restoring damaged landscapes and human communities. It is not just here in the Pacific Northwest; permaculture is a growing movement throughout the world.

Permaculture and Natural Farming

Permaculture has several characteristics that distinguish it from basic organic agriculture. It stresses the interconnectedness of life, is primarily a no-till system, emphasizes tree crops and perennial plants, and integrates the production of food and shelter with the social, economic, and political aspects of society. In many ways it *does* represent an improvement on basic organic techniques. But when carefully analyzed it becomes clear that permaculture, too, is a product of our modern way of thinking and therefore does not represent a clean break from other forms of modern farming.

The design process begins with careful observation of the patterns and interrelationships in nature. Then, in a process sometimes referred to as bio-mimicry, the designer imitates those patterns in the design. The practice of "observing nature," however, presupposes an observer and something that is being observed. The separation from nature is built into the process from the beginning, and what is produced is the designer's *impression* of nature.

The next step in the design process is an analysis of the elements and functions. Each element—say, a pear tree, chicken, or building—is examined and cataloged according to what its needs are and what it will provide to the other elements. This is essentially what science does. Nature, which is an indivisible whole, is split into individual parts, is analyzed, and then an attempt is made to put it back together. The reconstructed "whole" becomes a human invention, a simulation.

Ultimately, after all the information is collected and evaluated, the final design is produced. The designer is the creator and eventually the manager. Although the inspiration originally came from observing nature, the design is the work of the human intellect. The designer is firmly in *control.* I emphasized that word because it is perhaps the single most defining characteristic of modern culture.

So permaculture is based on the commonly shared beliefs and values of modern culture. It accepts people's alienation from nature without objection, or perhaps without noticing; analyzes the whole as bits and pieces; then tries to fit them together again. It relies on the human intellect every step of the way. There are many wonderful things about human consciousness, but it can never understand a reality that is inherently unknowable. The permaculture saying I quoted earlier, that "the possibilities in a permaculture design are limited only by the imagination of the designer," is meant to imply that the possibilities are limitless, but it also reveals a prideful overconfidence in human abilities. Why would anyone want to limit the possibilities to something as narrow and imperfect as the human mind?

Relying on the intellect moves permaculture outside the realm of natural farming. Most permaculturists, for example, would not object to spraying compost tea on their plants or using exotic soil amendments if they made the crop grow faster and created a higher yield. Using ultraviolet grow lamps is fine as long as they are powered by solar panels or a pedal-powered generator. They visit a natural pond ecosystem in a forest meadow and decide to re-create something like it in their design. The result is a high-tech aquaponics system using PVC pipes, pumps, bubblers, solar collectors, and a digital timer. It is not at all like what the designers first observed, or anything like what you would ever see on a stroll through the woods.

Permaculture also relies a great deal on information and scholarly research. The assumption is that if we gather enough information and analyze it with the aid of basic design principles we will eventually be able to figure out the best thing to do. David Holmgren, the co-founder of permaculture, explains, "Traditional agriculture was labour intensive, industrial agriculture is energy intensive, and permaculture-designed systems are information and design intensive."

Some advocates would like to see permaculture include more spirituality in its curriculum. Others believe that only provable, observable science should be allowed, and that "flaky metaphysics" should be purged. Mollison himself holds this position. In his book *Travels in Dreams*, Mollison writes, "As I have often been accused of lacking that set of credulity, mystification, modern myth and hogwash that passes today for New Age spirituality, I cheerfully plead guilty . . . permaculture is not biodynamics, nor does it deal in fairies, divas, elves, after-life apparitions or phenomena not verifiable by every person from their own experience, or making their own experiments."

The other camp believes there is more to the world than observable phenomena, that permaculture does not expressly forbid combining it with other disciplines, and that practices such as yoga, shamanism, and astrology have all passed scientific scrutiny. I was at one course where the instructor discussed his relationship with the fairies he had encountered in his garden and elsewhere. When asked if he actually believed in fairies, he said, "*I* do. Whether they are real or not . . . who knows? All I know is that they are real to me. They have made my life richer and more enjoyable, and have allowed me to practice permaculture more effectively than I would have been able to do otherwise."

Some followers believe that permaculture can change the direction of modern society by somehow subverting the mainstream toward a healthier path, but I am skeptical that any segment of our society can manage to do that when they adopt the same core assumptions as the culture they are hoping to change. Eventually they will be reintegrated into the mainstream because people will revert to their familiar ways of thinking and acting that were ingrained since infancy.

When permaculture first came to the United States it was portrayed as a decentralized, grassroots movement. It had an egalitarian, tribal feel that was quite appealing. Today, however, the trend is toward a more structured organization with a central "institute" and a panel of experts to regulate tighter standards for curricula and certification. This effort is largely promoted by those who think that working within universities, government agencies, and other mainstream organizations will allow them to reach a larger audience. They believe that if permaculture had a more professional feel it would make them more acceptable in the eyes of those institutions. It is actually the first step toward *becoming* those institutions. A more structured organization leads to more centralized control, and with that comes the inevitable struggle to gain and maintain the upper hand. All permaculture needs to do to avoid this is to stay true to its egalitarian, decentralized roots. That means building strong, resilient, local cooperative communities outside the stifling influence of orthodoxy.

Although I have been pointing out what I consider to be shortcomings in the permaculture approach, I am not saying that permaculture has not had a positive influence. It has. Millions of trees have been planted over the past forty years that would not have been planted otherwise. Skills like seed collecting, food preservation, natural architecture and building, growing and using medicinal plants, foraging for wild plants, and how to use appropriate technology effectively have been revived, helping to conserve valuable knowledge and resources. Cooperative permaculture communities have grown up all over the world, providing an entryway for many people to come in contact with the natural world perhaps for the first time. But I believe it should be seen as just that, an entryway, not an end in itself. Ultimately, permaculture, too, must be transcended in favor of a broader vision that holds serving nature as its highest priority. If nature is the perfect model, why not let nature do the design?

Shortly after *The One-Straw Revolution* was published a woman mentioned to me that the message in the book was somehow familiar, as if it had been in her subconscious for some time, but this was the first time she heard it expressed in words. I think this is because Mr. Fukuoka is speaking to us about an earlier time when people were fully engaged with nature. We still remember, but it has become a distant memory. Everything was given to us then as from a continuously flowing spring.

Natural farming is a timeless understanding that is current and applicable in any age. It uses nature as its standard, and nature's truth does not change. In developing his farming method Mr. Fukuoka found his way back to the source by "whittling away unnecessary agricultural practices." When he finally succeeded in eliminating all of them, he was left with "nothing." Only then did nature's true form appear, along with its full reward. When you walked through Mr. Fukuoka's orchard you saw nature, fresh and clear.

Other agricultural systems reflect our current isolation from nature. What you see is not nature but the product of human will. Industrial agriculture defiantly and unapologetically displays modern culture's mission of complete domination. Organic agriculture and permaculture are not as harsh, but the evidence of human control persists. I realize that there is a certain satisfaction and pride that comes with creating a plan and then seeing it take physical form, especially when you have patterned the design after natural systems, but it will always be a simulation. What you see is yourself, not nature's unique and wondrous expression.

Mr. Fukuoka was often asked by farmers here in the United States about the practicality of mixing his method with other organic techniques. He politely said that it was not possible. Recently I received a similar question from a grain farmer who lives in the Willamette Valley. "I am having a terrible time controlling weeds after I do mechanical tillage," he wrote. "Can natural farming help me with that?" The short answer, of course, is no. What this farmer was *really* asking was, "How can natural farming allow me keep doing what I have been doing but without the side effects?" He wanted to pile on more techniques, hoping for a magic bullet. This approach is the opposite of natural farming. Modern farming wants to keep doing things; natural farming stops doing them.

When considering how to approach a particular piece of land, Mr. Fukuoka encouraged us to first ask the question, "What does the land need?" not "What can nature provide." In most cases the land has been damaged in some way, so the first step is to give nature the tools it needs to fully express its predilection toward sustaining life. That means to stop doing the things that caused the damage in the first place, then reviving the soil, and reestablishing diversity. More often than not, planting trees and establishing a permanent soil-building ground cover is a good way to begin. The farmer gives the land what it needs and eventually the land is able to reciprocate in kind.

Farming the way Mr. Fukuoka suggests does not come naturally to most people, because they have come to believe they have to be in control to guarantee a successful outcome. It *does* involve a leap of faith to leave that way of thinking behind. It is as if you are standing on the edge of a cliff and Mr. Fukuoka is there encouraging you to step off, saying, "Don't worry, everything will be fine." When you do step off it turns out not to have been a cliff at all but a step into a new and magnificent world. That reality is accessible to anyone at any time. The world does not change; reality never changes. It is only our perception that changes.

CHAPTER 8

Without Natural People, There Can Be No Natural Farming

MR. FUKUOKA SUGGESTED THAT FOR SOMEONE to clearly assess natural farming they first need to come in direct contact with nature as it truly is; then they can decide for themselves whether or not to follow that path. That is easier said than done, of course, but there are certain steps we can take to align ourselves with the natural order, both inside and out. In so doing we put ourselves in a position where seeing the world the way he did becomes possible. It is mainly a process of removing obstacles, letting go of misconceptions, and living a simple life that is close to the heart of nature.

Some people use field trials, scientific examination, and statistical analysis to test the methods of natural farming, but these trials are typically performed in unnatural environments that have been created specifically to accommodate modern farming methods. Natural farming works best when conditions are as natural as possible because it relies on nature's unique ability to produce abundance by itself. It will always come up short when tested under unnatural conditions.

Others, even those who are predisposed to accepting the approach of natural farming, have difficulty evaluating the efficacy of the philosophy because their interior landscapes have also been groomed to serve the objectives of modern society. The antidote is to examine our own thoughts and determine those that have been placed there by our culture. A good way to do that is by using the same method Mr. Fukuoka used to clarify his thinking with regard to agriculture. When Mr. Fukuoka developed his farming method he began by considering the most basic agriculture practices, such as plowing, pruning, making compost, and flooding the rice fields. He asked himself, "Are these practices really necessary?" Eventually he decided that none of them was, so he stopped doing them.

In examining our thoughts we need to ask ourselves a similar question: "Is this a universal truth or is it a concept unique to our modern culture?" If it is unique to modern culture it should be discarded. Some of these ideas might include, "Society needs to keep growing so we can accomplish great things and make continual progress. It's better to do something than nothing. We need to understand how nature works so we can extract what we need from it in order to meet the needs and demands of our growing population. We need industrial agriculture to feed that ever-growing population. The primitive people who came before us lived a pitiful and impoverished existence."

Some thoughts may be more personal, for example, "I need to have a career and accumulate material possessions to consider myself a success. The more I know, the better off I'll be. I need to have a plan for my life or things will turn out poorly. I need to discover the true meaning of life." Mr. Fukuoka denied that any of these statements was true because they are all the product of a culture that did not use nature as its touchstone for truth.

It is hard to leave all that behind. After all, these are powerful values and beliefs that are shared widely. They can define our sense of self and our sense of purpose. Abandoning these beliefs can be frightening, seem risky, and—in a culture where they are considered to be unassailable truths—may seem extreme. But I think you will find that ultimately it is much easier to unburden yourself from all that misinformation than it is to carry it around with you. In the end, it will be replaced by other things that are infinitely more fulfilling.

The next step is to examine how we live with the idea of eliminating non-essential activities and material clutter. That means reducing material possessions, limiting purchases of consumer goods and services, making thoughtful and environmentally sound decisions when it comes to transportation, housing, food, and entertainment, and creating as little waste as possible. This not only makes life easier, but also gives us a sense of independence and personal empowerment. It is moving the economics of our personal lives from the economics of scarcity to the economics of abundance. It is taking from the world only what we need and no more.

This necessarily involves uncoupling ourselves from the belief that accomplishing great things and amassing material goods will lead to a more comfortable and fulfilling life. It is easy to see the damage humanity's preoccupation with progress has caused: the exhaustion of resources, the contamination of the air, soil, and water, the extinction of species, and the confusion and demoralization of human society. What is not as apparent is the damage we cause ourselves by pushing ourselves to accomplish great things in our personal lives.

We hear from childhood that we need to work hard in school so we can have a successful career, preferably one that will make us a lot of money, but should material gain be the benchmark for personal happiness? For many, living materialistically turns into a life of insatiable desire, an addiction that robs them of all hope for achieving fulfillment in life and the sense of empathy for others. It also anchors them securely in the world of relativity where there is no freedom, only continuous confusion and conflict.

Wouldn't it be better to calmly go about our daily routine, live simply, eat and sleep well, and just enjoy being alive? Living simply leaves room to enjoy the richness of each day. It puts the emphasis on the value of experiences rather than belongings. It aligns our lives with what we believe in, and that leads to a more positive and more confident attitude. Wearing simple clothing, eating simple food, and living a humble, ordinary life elevates the human spirit by bringing us closer to the source of life.

For many, this is where the path ends. They are satisfied to set up comfortable homesteads that are patterned after nature where they can live healthy self-reliant lives, perhaps in a remote area that is completely off the grid. Others rearrange their lives so they can live this way even while living in the city. Living at a slower pace, tending a small garden, and living a simple existence centered around the hearth is ideal for them. They are helping to put the brakes on the senseless destruction of nature by working to heal the land and live responsibly.

The Final Leg of the Journey

This is all well and good, but if we stop here we will still be left in the world of relative thought, believing that we exist as individual entities and therefore separate from Creation. We haven't made it to natural farming yet. To get there we need to keep getting rid of more stuff.

To get this far most people have relied on various aids to help them restore order in their lives. These practices have been useful for healing and for providing a more balanced outlook. They are different for each person, but some might include studying one of the spiritual arts, macrobiotics, astrology, permaculture, or biodynamics, to name a few. As we continue on, however, the structure of these practices needs to be cast aside. Their mission is complete. We should be grateful for what they have given us, but from here we will go on without them. Whatever "isms," philosophies, or religious dogmas we have chosen to adopt as our personal standard for truth also need to be discarded. They are only useful in the relative world where no absolute standard for truth exists.

By now we are beginning to feel light and free, but it is starting to get a bit scary. We are traveling without a map or a plan, and the intellectual reasoning we have grown accustomed to using to set our course is not helpful anymore because we are no longer in the familiar world of the intellect. It is a place we have not been since childhood but seems strangely familiar.

Actually, traveling without a plan is the best plan of all because it allows for good things to happen unpredictably. Magically, you are led to a place you never expected to be but truly belong. If Mr. Fukuoka had started with a predetermined idea of what his natural farm would be like it never would have taken the magnificent form that it did. If I had had an inflexible itinerary when I set out for Asia I never would have found my way to the countryside and eventually to Mr. Fukuoka's farm. When things are left unplanned, everything is possible. Think of all the things you found to do when you were a child and were allowed to play freely in nature. With a preset structure those wonderful fantasies and unexpected discoveries would never have happened.

Katrina Blair, in her wonderful book *The Wild Wisdom of Weeds*, describes an experience she had when she was eleven years old while camping with her family at a high mountain lake north of Durango, Colorado. She paddled by herself to the far side of the lake on a pool mattress and walked ashore. As she recalls

> *I sat down in and among all the wild mountain plants and became filled with an intense and completely joyful energy that I had never experienced before in such potency. It was a feeling of total euphoria! The plants and the surrounding environment gave me a strong message in that moment, saying "You are home" and "You are going to live your life with us." The essence of this feeling still speaks to me today in my core, reminding me that I am at home here on our wild Earth.*[1]

It is not possible to *will* an experience like this to happen, but if nature is approached with a clear mind and an open heart its true form will appear without conscious effort. The experience Mr. Fukuoka had when he was a young man also came unexpectedly, without effort on his part. The more we grasp at something and try to make it our own, the more it eludes us. All we can do is put ourselves in a position where such an experience is possible. The door will open when you have forgotten to think about it any longer. It is like looking at one of those op art drawings of a person's face. After a while, it seems to switch to a profile view. The picture does not change and there is no way to *make* yourself see the picture the other way, it just happens.

That's the feeling when you unexpectedly see the world as it is for the first time. It's like, "Wait a minute, what just happened." The first time I had an experience of this kind I was hiking in a redwood forest on the north coast of California. I stopped for a moment and sat near a small creek. When I looked up everything was different. It looked the same, but for some reason I was looking in instead of out. Everything seemed to be perfectly in order as it had been since the beginning of time, and there was a profound sense of peace that took my breath away. I celebrated by hugging a tree, saying hello to a huckleberry bush, and enjoying the smell of a handful of duff I picked up from the forest floor. A squirrel running along a tree branch, the sound of the water, the sunlight streaming through the leaves—everything was a revelation, a miracle, and there was no end to it! I felt a rush of joy and compassion. It was as if I had become part of something greater than myself, something wonderful.

Until that time I had been gradually simplifying my life with the vague goal of getting rid of everything until there was nothing left, a kind of do-nothing/have-nothing program. I examined my thinking, pared material needs and possessions to a minimum, cleaned up my diet, and generally lived a modest life. These were conscious decisions. I had a destination in mind. I didn't think for one minute that I would ever get there, but I was feeling lighter and closer to the earth all the time and that was reward enough. I never dreamed something like this would happen.

The experience I had that day contained an irony that made me laugh out loud at the time. Besides the effort to simplify and achieve greater clarity in my life I was also reminding myself to not plan and not grasp after things, but I was always aware that it was "I" who was doing it. When the final possession, my ego or sense of self, dropped away, "I" had nothing to do with it. It was a gift, pure and simple, and I was filled with gratitude. I was finally free of my "self." It is impossible to describe such a joyful feeling, but others have referred to it as "joining with cosmic consciousness" or, as Sensei put it, becoming part of "big life." I had reached my destination. *Everything* was gone, but something wonderful had rushed in to take its place. I felt like nature was flowing through me in the same way it flowed through Mr. Fukuoka's farm.

Here's the next irony: This destination turned out to be the beginning of an entirely new journey. My heart was filled with the song of Creation, but now what? I decided to turn my life into a natural garden in which nature had free rein. I still live a rather ordinary life, but it is filled with gratitude and a sense of complete freedom.

When you tune yourself to the "great unfolding," miracles happen. You don't have to be a farmer to experience this. Living a simple life dedicated to service reminds others that there *is* another way to live, one that is infinitely more satisfying than struggling in the pointless pursuits of the relative world. The one-straw revolution is about remembering who we are so we can live freely, joyously, and responsibly in the world. As Mr. Fukuoka pointed out, "Without natural people there can be no natural farming."

APPENDIX A

Creation, by Harry Roberts[1]

IN THE BEGINNING THERE WAS JUST SPIRIT. Spirit is made of Spirit-stuff. It is not a solid substance or a gas, just Spirit-stuff, which is different from physical-stuff. It is a force, an energy, which creates and exists yet has no form or dimensions. It makes all things of itself yet is itself neither diminished nor consumed. It is timeless and limitless.

Each thing is given its own spirit while behind each spirit is the Spirit of Creation, or just Spirit. So there is only one Spirit and all things are as one. I call it "Spirit" or "Creation," but those are just words; it is everything.

To understand this force one has simply to accept it and to accept one's own position in it. It is impossible to tell somebody that if you are in complete acceptance of Creation you will not have even a concept of the feeling of fear, but this is true.

When the white man came the Indians tried to tell them about Creation and the white man called it, "The Great Spirit." However, this was conceived of as an overgrown personification of a man who lived somewhere in the sky, like Jehovah, and this was far from what was meant.

An ant is physically and mentally limited; it cannot conceive of the earth. Man is in the same position in the face of Creation. To personify Creation in the form of a man or a god is juvenile. A man is one who can stand alone and does not need a god to blame things on. But the facts of life are that very few people are strong enough in spirit to stand alone; they need crutches to support themselves. (I suspect that these are "the lame and the halt and the blind" that Jesus miraculously cured so that they could see and throw away their crutches.)

Of course, if there is no personification of God, there need be no devil, either.

Since the Yurok had no personified God they had no name for God. The European insisted on having a name so some Yuroks said "Wohpekumew," who put everything in order in the beginning, was God.* But others gave a word that meant "creation"[*ki 'wes'onah*] or "to exist," "to be" or "that is." Sometimes people used a word like *nahkwok'* [it does] meaning by it, "You see how it is like but there are no words to say it." The description of Creation was usually given by an expressive opening gesture of the hands. Still, I call it "Creation."

Creation is everything. The old-timers felt that an understanding of Creation could be achieved through the study of beauty.** The world was seen as beautiful and their prayer was, "May you walk in beauty." Today there is art but much art is ugly because there are no spaces, and spaces are where the spirits walk in the beauty of Creation.

Before, when a man made communion with Creation so that he could walk in beauty he stood forth on a mountain top and opened his hands and held his arms wide and looked full into the breaking dawn and let the Spirit of Creation flow into him. He didn't even wear moccasins or a necklace lest some portion of him should be shielded from the light of Creation. To these men it seemed very strange to see their grandchildren clasp their hands together to hide something and bow their heads and close their eyes to hide their spirit when they were hidden behind walls and call that prayer.

But these were real *men*. I don't mean big bruisers stomping around, but complete people. We say "man," but some of them have been women. These people became complete by studying very hard, training hard from their early life on.

---

* In Yurok mythology, Wohpekumew is one of the two (quite fallible) great trickster-creators of the world. The other is Pulekuwerek. Neither made the earth; they only put it into its now familiar order and rhythm.

** Harry rendered the Yurok *mrwrsrgerh* as "beauty." In Yurok it is a verb, "to be beautiful," in the sense of being pure, unpolluted, able to make medicine and to take part in rituals.

APPENDIX B

Akwesasne Notes Review of *The One-Straw Revolution* (1978)[1]

For countless generations, since the introduction of agriculture to the Eastern Woodlands, the Native people of North America raised their crops without the use of the plow or other facets of European agriculture. Surprisingly, although there was little machinery, there was also little (by historical agricultural standards) labor—the Seneca, for example, apparently hoed their fields only once between planting and harvesting.

Agriculture, in former times, was accomplished with little more than a sharp stick and a fairly extensive oral tradition of know-how. The forests of Colonial and pre-Colonial times provided an abundant harvest of their own—fish, game, berries, herbs, roots and so forth. The Natural World, intact, is an abundant supplier of local human needs.

The Hau de no sau nee, or Six Nations people, are among those groups which have retained a great deal of memory of agriculture in early times. The Hau de no sau nee could be conceived of as a people divided into two communities—a community of men and a community of women. History and tradition both record that the jobs associated with agriculture under the traditional economy were primarily the province of the community of women. Traditional Native agriculture was very different from European agriculture, ancient and modern.

In ancient times, trees were girdled to cause the leaves to fall (and to provide a standing source of dry firewood), and gardens were laid among the girdled trees. Seeds were planted in groups on small hills in the forest-enriched earth, fertilized when possible with fish scraps, tended with little more than a stick and a hoe, and left to grow. The favorite crops were corn, beans and squash.

Some of the best accounts of the results come from military journals, especially journals from the officers of expeditions against the Indians such as the Clinton-Sullivan campaign during the American Revolution. American soldiers reported finding extensive fields of corn, beans, and squash, and large orchards in the Indian country. Their reports claim that they destroyed millions of bushels of grain in 1779 in the region of the Finger Lakes and the Genessee River of Central New York. The eyewitness accounts of the agricultural production of those regions will astound the unenlightened.

The day when Western agricultural methods were introduced among the Seneca (one of the Six Nations) is recounted in the historical record. Well-meaning Americans (primarily Quakers) approached the Senecas along the Allegany River in the 1790s with offers to teach new techniques of agriculture which involved draft animals and the iron plow. An experiment was conducted whereby two fields were planted—one field utilizing the traditional hilling technique, and the other cleared and plowed according to the European custom. It is reported that at the first harvest, the plowed field produced a slightly larger harvest, and that thereafter the Senecas adapted the technique wholeheartedly. Over the next century, the traditional technology was largely abandoned.

Today, a message comes from a spiritual person of Japan which calls for a serious and intensive return to the traditional agricultural technology. The book which contains this message was written by Masanobu Fukuoka and is entitled *The One Straw Revolution—An Introduction to Natural Farming*. It is a book which Native and Natural people would be wise to read carefully. Mr. Fukuoka began his adult life as an agricultural scientist who, while still in his twenties, began to question assumptions about large-scale agriculture, and even about agriculture as it has been practiced in Japan for the past 400 years. His reservations about those practices, combined with a strong spiritual vision of the world, have led him to successfully develop an agricultural technique which requires no plowing, no pesticides, no herbicides, no weeding, no chemical fertilizer; he doesn't even use organic compost. He calls the process "natural farming." Although the process he advocates arises in southern Japan, and it utilizes crops appropriate to Japanese climate and culture, the philosophy and practice of the technique is amazingly close to that of Native peoples prior to the introduction of European agriculture. To be sure, there are critical innovations—the use of straw and the conscious seeding of legumes such as clover and alfalfa—but there are striking similarities between his techniques and the traditional Native ways of doing things.

European agricultural technology as it was transplanted to the Americas (and since then, to the world,) has always had its drawbacks. European agriculture is distinguished by the process of clearing the land and turning the soil and then setting forth to accomplish the biological simplification of the land until only one life form remains on the field. It is the European farmer's objective that the only thing left standing in a field of cabages are the cabages [sic]. That process has led to a lot of problems for the farmer.

Plowing the land and planting it to one crop decreases rapidly the fertility of the soil, requiring that the land be refertilized with animal dung, decomposed plant and animal wastes, or chemical fertilizers. Or replanted to a soil-enriching legume, such as clover. The decreased fertility of the soil can be assumed to lead to weakened plants which are more susceptible to disease and parasite infection, and there is much discussion in modern circles to the effect that the plants contain less nutrients for the people who eat them than do plants raised in naturally fertile soil.

Mr. Fukuoka argues that the European mental process, as applied to agriculture, has been trying to find solutions to the problems of plant production on a one-at-a-time basis, rather than seeking the root causes. His arguments are powerfully persuasive, profoundly radical, and spiritually stimulating. He states that the first mistake is made when the land is plowed. That position would probably be dismissed as the mindless ramblings of a hopeless romantic except for the facts that he been practicing what he preaches for more than twenty years and has produced yields comparable to the yields obtained under the most modern chemical techniques, and his position enjoys some historical verification.

The introduction of European agriculture in the Seneca country in the 1790s set into motion a series of processes which are well worth reviewing. According to the accounts, turning the land with a plow did produce an increased crop yield that first year. It is predictable that the particular piece of land was already naturally fertilized, and by definition, had not been worked for a number of years. But once the land was turned, there were introduced a new set of needs which people rarely talk much about.

Cleared land agriculture required animal power, and horses or oxen must be sheltered, watered, and fed. It came to pass that a lot of land had to be cleared for that purpose—hay had to be planted, a great deal more grain needed to be harvested to provide food for the work animals. And it also became necessary to use domestically-produced animal fertilizer, which meant that manure had to be gathered and spread on the fields. Males, who had traditionally been involved primarily in hunting and fishing, now became workers on the homestead not, as has been suggested, because the work was too heavy for women, but because there was such an enormous increase in the amount of work which needed to be done. There were social factors at play in all this, to be sure, but it is undeniable that the amount of work in agricultural production increased several fold following the introduction of this way of doing things. It involved a lot of hard work.

Mr. Fukuoka argues that a lot of it is unnecessary work. He suggests that wnever [sic] Humanity interferes with Nature (as when a field is plowed) . . . things start to go wrong. Once you turn the land, then you must use fertilizer and you need to fight the weeds and there arise all kinds of problems with insects, and plant diseases. The European solutions to these problems—chemical fertilizers, pesticides, herbicides, and complicated expensive machinery—cost great amounts of money, create pollution, and produce debased food. The chemical solutions continue to deplete the soil and cause the destruction of plant and animal life, while the increased mechanization serves to force people away from an agricultural life while enriching industrialists. We are accustomed to being told that mechanization leads to a more affluent and easier life. Mr. Fukuoka suggests that instead, we live in a fool's paradise (not his words). Natural farming technologies, he states, produce the same yields as do chemical and machine-intensive technologies, are much less destructive to the environment, require less human labor, and enable people to scale agriculture down to human dimensions. In addition to that, one should add, the natural ways produce much better food, and create broad possibilities for a more human lifestyle.

The powerful aspect of Mr. Fukuoka's message does not involve natural farming technology, however. His message is timeless and speaks to the nature of human existence. He is a Natural World philosopher, a man with an enormous appreciation of the forces of Creation, and one who understands the potential (and historical) follies of the human mind. The book is more of a philosophical treatise than a technical manual. He explains the basics of his technique, which involves the expert use of cover crops and the reintroduction of complex biology into agriculture, but he is at his best describing his philosophy of Nature. For the most part, his message could have been spoken by a Lakota, a Seneca, or a Zuni traditionalist. That this specific message comes from Japan is a powerful indicator that Natural people have a strong common bond throughout the world.

APPENDIX C

Mother Earth News Interview with Masanobu Fukuoka (1982)[1]

MASANOBU FUKUOKA, WITH HIS GRIZZLED WHITE BEARD, subdued voice, and traditional Oriental working clothes, may not seem like an apt prototype of a successful innovative farmer. Nor does it, at first glance, appear possible that his rice fields—riotous jungles of tangled weeds, clover, and grain—are among the most productive pieces of land in Japan. But that's all part of the paradox that surrounds this man and his method of natural farming.

On a mountain overlooking Matsuyama Bay on the southern Japanese island of Shikoku, Fukuoka-san (*san* is the traditional Japanese form of respectful address) has—since the end of World War II—raised rice, winter grain, and citrus crops . . . using practices that some people might consider backward (or even foolish!). Yet his acres consistently produce harvests that equal or surpass those of his neighbors who use labor-intensive, chemical-dependent methods. Fukuoka's system of farming is amazing not only for its yields, but also for the fact that he has not plowed his fields for more than 30 years! Nor does he use prepared fertilizer—not even compost—on his land, or weed his rows, or flood his rice paddies.

Through painstaking experimentation, you see, this Japanese grower has come up with a method of agriculture that reflects the deep affinity he feels with nature. He believes that by expanding our intellect beyond the traditional confines of scientific knowledge—and by trusting the inherent wisdom of life processes—we can learn all we need to know about growing food crops. A farmer, he says, should carefully watch the cycles of nature and then work with those patterns, rather than try to conquer and "tame" them.

In keeping with that philosophy, Fukuoka-san's fields display the diversity and plant succession that is a natural part of any ecosystem. In the spring, he sows rice amidst his winter grain . . . then, late in the year, casts grain seed among the maturing rice plants. A ground cover of clover and straw underlies the crops, deterring weeds and enriching the soil. In addition, the master gardener grows vegetables "wild" beneath the unpruned trees in his mountainside orchard. Naturally, such unconventional plots might look positively disastrous to traditional agronomists, but as Fukuoka points out to skeptical visitors, "The proof of my techniques is ripening right before your eyes!"

For many years, the Oriental gentleman's unique ideas were known only to a few individuals in his own country. In 1975, however, he wrote a book entitled *The One-Straw Revolution*, which was later published in the United States. Since then, he has been in great demand by groups eager to know more about this strange "new" attitude toward farming. In 1979 Fukuoka-san undertook an extensive tour of the United States . . . and while he was in Amherst, Massachusetts for a series of university lectures, he talked for several hours with Larry Korn, a student of natural farming methods and the editor of *The One-Straw Revolution*. Their conversation was conducted entirely in Japanese and later translated into the edited version printed here.

Incidentally, if you're puzzled by several instances of apparent contradiction in the following comments, consider that Fukuoka—like the Oriental philosophers who deliberately present students with what seem to be illogical statements or paradoxes—is perhaps trying to help people break habitual patterns of thought and develop new perceptions. And because his natural farming does demand such an unaccustomed mode of thinking, Fukuoka-san warns that it is not for the timid or the lazy: "My method completely contradicts modern agricultural techniques. It throws scientific knowledge and traditional farming know-how right out the window." What's left in the wake of that revolutionary (and sometimes admittedly befuddling) upheaval, however, should excite—and challenge—anyone who'd like to see a simpler, more natural form of agriculture take root.

Mother Earth News: I notice that you're drawing, Fukuoka-san ... what will the picture be?

Fukuoka: It's a sketch of a mountain scene, and there's a poem with it:

Deep in the mountains, a gentle soul asks,
For whom do the wildflowers bloom?
For foxes and raccoons,
Who know the pine winds and
The spirit of the valley stream.

Mother Earth News: Can you explain what you mean by that verse?

Fukuoka: Well, there are many ways of defining this "gentle soul." It could be a person ... a flower ... a tree ... or even the grass. And if one could ask this soul why it lived all alone, deep in the mountains, it would answer, "I am not living here for anybody's sake. Just to listen to the fox and the raccoon, to talk to them and be with them ... that is why I am living here."

Mother Earth News: Are you the figure I see in the drawing?

Fukuoka: I'd like it to be me!

Mother Earth News: Well, it's certainly evident from your artwork and from your approach to farming that you value having a close relationship with nature. Were you raised in a rural setting?

Fukuoka: Yes, I was an ordinary country boy, born in a simple country house. My father—who served as the leader of our small village—was a landowner and farmer. I grew up just as the other local children did ... going to school and helping my parents and neighbors in the rice fields.

Mother Earth News: Did you begin farming as soon as you had finished school?

Fukuoka: No, I first went to a special technical institute to study microbiology and plant pathology. Then I moved to Yokohama to become a quarantine officer at the Agricultural Customs Office. My job was to inspect, and experiment with, Japanese mandarin oranges and American oranges. I learned a lot there about the weaknesses and diseases of different plants . . . and greatly enjoyed my laboratory work. However, at the age of 25, I underwent a change of heart—and mind—that caused my life to be completely different from that time on.

Mother Earth News: Tell me about it.

Fukuoka: Well, like many young people, I was having very large, ponderous thoughts about life . . . and my musings led to a lot of skepticism about the human condition. To add to my doubts, I became so ill during that period that there was, for a while, a question of whether or not I would pull through.

Following my eventual recovery, I spent many sleepless nights wandering the streets. The morning after one such episode—when it seemed as though everything were about to explode in my brain—a flash of insight came to me. I suddenly felt that all human existence is meaningless and of no intrinsic value. Humanity knows nothing of real worth at all, I decided, and every action we take is just a futile, empty effort. I also saw that nature is ideally arranged and abundant just as it is . . . therefore, I was sure that we should work in cooperation with the natural processes, rather than try to "improve" on them by conquest.

I know all this may sound preposterous, but whenever I try to put those thoughts into words, they seem to sound that way. The revelation wasn't something that can be easily explained to another person.

Mother Earth News: Why not?

Fukuoka: Anyone who's had an experience similar to mine will understand instinctively . . . but there's nothing I can say to help those people who don't have this understanding or aren't even looking for it. For example, do you think there's such a thing as a ghost? Have you ever seen a ghost? [With a smile, he points over the interviewer's shoulder.] Didn't you just see that one? People who've never seen a ghost usually can't believe in them. Those who have had such an experience, though, totally believe in the phenomenon . . . so there's no need to convince them.

Mother Earth News: How did this change in thinking affect your life?

Fukuoka: I immediately quit my job at the Customs Office. Then I spent the next year or two traveling around the country, talking with people and trying many new experiences. Sometimes I camped in the mountains and sometimes near hot springs. Whenever I was in a city, I would sleep in temples or parks . . . and when I was in the country, I stayed at farmers' homes and worked in their fields with them. I actually started my wanderings with the intention of spreading my new understanding throughout the whole country . . . but whenever I spoke about the meaninglessness of human existence, nobody was interested in what I had to say. I was ignored as an eccentric. So I finally decided that in order to help people understand my theories, I'd have to demonstrate them in some concrete and practical way. I also needed to do that, of course, to convince myself that I was right.

Since I believe that farming is the most worthy of all occupations, I decided to return to my native village and become a farmer. I wanted to see whether I could apply my theory of the uselessness of human knowledge to agriculture . . . so that if people didn't understand my words, I could take them out to the fields and show them the truth of these ideas.

Mother Earth News: And you've been farming ever since?

Fukuoka: Almost. During the Second World War, I was sent to work at the Agricultural Experimental Station at Kochi, where I had to fall back upon my scientific training. After the war was over, though, I joyfully returned to the mountains and resumed my life as a farmer.

Mother Earth News: How much land did you start with?

Fukuoka: After the war there was a massive land reform in Japan—called the Nochi-kaiho—in which large landowners like my father lost most of their holdings. My father died soon after that, and I was left with one small rice paddy about a quarter acre in size.

Mother Earth News: Did you begin practicing natural farming right away?

Fukuoka: I had started experimenting in some of my father's mandarin orange orchards even before the war. I believed that—in order to let nature take its course—the trees should grow totally without intervention on my part, so I didn't spray or prune or fertilize . . . I didn't do anything. And, of course, much of the orchard was destroyed by insects and disease.

The problem, you see, was that I hadn't been practicing natural agriculture, but rather what you might call lazy agriculture! I was totally uninvolved, leaving the job entirely to nature and expecting that everything would turn out well in the end. But I was wrong. Those young trees had been domesticated, planted, pruned, and tended by human beings. The trees had been made slaves to humans, so they couldn't survive when the artificial support provided by farmers was suddenly removed.

Mother Earth News: Then successful natural farming is not simply a do-nothing technique?

Fukuoka: No, it actually involves a process of bringing your mind as closely in line as possible with the natural functioning of the environment. However, you have to be careful: This method does not mean that we should suddenly throw away all horticultural knowledge that we already have. That course of action is simply abandonment, because it ignores the cycle of dependence that humans have imposed upon an altered ecosystem. If a farmer does abandon his "tame" fields completely to nature, mistakes and destruction are inevitable. The real path to natural farming requires that a person know what unaltered nature is, so that he or she can instinctively understand what needs to be done—and what must not be done—to work in harmony with its processes.

Mother Earth News: That attitude certainly denies the "manipulate and control" foundation of modern agriculture. How did you progress from your traditional training to such an unusual concept of farming?

Fukuoka: During my youth I had seen all the farmers in the village grow rice by transplanting their seedlings into a flooded paddy . . . but I eventually realized that that isn't the way rice grows on its own. So I put aside my knowledge of traditional agricultural methods and simply watched the natural rice cycle. In its wild state, rice matures over the summer. In the autumn the leaves wither, and the plant bends over to drop its seeds onto the earth. After the snow melts in the spring, those seeds germinate, and the cycle begins again. In other words, the rice kernels fall on unplowed soil, sprout, and grow by themselves.

After observing this natural process, I came to view the transplanting/flooded field routine as totally unnatural. I also guessed that the common practices of fertilizing a field with prepared compost, plowing it, and weeding it clean were totally unnecessary. So all my research since then has been in the direction of not doing this or that. These 30 years of practice have taught me that many farmers would have been better off doing almost nothing at all!

People often think, in their arrogance and ignorance, that nature needs their assistance to carry on. Well, the truth is that nature actually does much better without such "help" from humans. Once a field is healthy and working on its own, natural—or "non-interference"—agriculture becomes a real possibility. However, as my citrus grove demonstrated, such a condition can't be initiated suddenly. In Japan and other agricultural countries, the land has been plowed by machines for decades . . . and before that it was turned by cows and horses. In fields such as those, you wouldn't have very good results in the beginning if you simply stopped cultivating the earth and adopted a do-nothing attitude. The soil must first be allowed to rehabilitate itself. Fertility can then be maintained by surface mulch and straw that break down into the soil.

Mother Earth News: For folks who may be unfamiliar with your book, *The One-Straw Revolution*, let's review the basic practices you follow in your natural system of growing grain, vegetables, and citrus.

Fukuoka: First of all, I operate under four firm principles. The first is NO TILLING . . . that is, no turning or plowing of the soil. Instead, I let the earth cultivate itself by means of the penetration of plant roots and the digging activity of microorganisms, earthworms, and small animals.

The second rule is NO CHEMICAL FERTILIZER OR PREPARED COMPOST. I've found that you can actually drain the soil of essential nutrients by careless use of such dressings. Left alone, the earth maintains its own fertility, in accordance with the orderly cycle of plant and animal life.

The third guideline I follow is NO WEEDING, either by cultivation or by herbicides. Weeds play an important part in building soil fertility and in balancing the biological community . . . so I make it a practice to control—rather than eliminate—the weeds in my fields. Straw mulch, a ground cover of white clover interplanted with the crops, and temporary flooding all provide effective weed control in my fields.

The final principle of natural farming is NO PESTICIDES. As I've emphasized before, nature is in perfect balance when left alone. Of course, harmful insects and diseases are always present, but normally not to such an extent that poisonous chemicals are required to correct the situation. The only sensible approach to disease and insect control, I think, is to grow sturdy crops in a healthy environment.

As far as my planting program goes, I simply broadcast rye and barley seed on separate fields in the fall . . . while the rice in those areas is still standing. A few weeks after that I harvest the rice, and then spread its straw back over the fields as mulch. The two winter grains are usually cut about the 20th of May . . . but two weeks or so before those crops have fully matured, I broadcast rice seed right over them. After the rye and barley have been harvested and threshed, I spread their straw back over the field to protect the rice seedlings. I also grow white clover and weeds in these same fields. The legume is sown among the rice plants in early fall. And the weeds I don't have to worry about . . . they reseed themselves quite easily!

In a 1¼-acre field like mine, one or two people can do all the work of growing rice and winter grain in a matter of a few days, without keeping the field flooded all season . . . without using compost, fertilizer, herbicides, or other chemicals . . . and without plowing one inch of the field. It seems unlikely to me that there could be a simpler way of raising grain.

As for citrus, I grow several varieties on the hillsides near my home. As I told you, I started natural farming after the war with just one small plot, but gradually I acquired additional acreage by taking over surrounding pieces of abandoned land and caring for them by hand. First, I had to recondition that red clay soil by planting clover as a ground cover and allowing the weeds to return. I also introduced a few hardy vegetables—such as the Japanese daikon radish—and allowed the natural predators to take care of insect pests. As a result of that thick weed/clover cover, the surface layer of the orchard soil has become over the past 30 years loose, dark-colored, and rich with earthworms and organic matter. In my orchard there are now pines and cedar trees, a few pear trees, persimmons, loquats, Japanese cherries, and many other native varieties growing among the citrus trees. I also have the nitrogen-fixing acacia, which helps to enrich the soil deep in the ground. So by raising tall trees for windbreaks, citrus underneath, and a green manure cover down on the surface, I have found a way to take it easy and let the orchard manage itself.

Mother Earth News: Don't you also grow vegetables in a kitchen garden?

Fukuoka: Actually, I raise such produce, in a semi-wild manner, among the weeds all over the mountain. In my orchard alone I grow burdock, cabbage, tomatoes, carrots, mustard, beans, turnips, and many other kinds of herbs and vegetables. The aim of this method of cultivation is to grow crops as naturally as possible on land that might otherwise be unused. If you try to garden using "improved" high-yield techniques, your attempt will often end in failure as a result of infestation or disease. But if various kinds of herbs and other food crops are mixed together and grown among the natural vegetation, pest damage will be so low you won't have to use sprays, or even pick bugs off by hand.

To plant my vegetable crops, I simply cut a swath in the weed cover and put out the seeds. There's no need to top them with soil ... I just lay the cut plants back over them as a natural mulch. Usually the resurgent weeds have to be trimmed back two or three times afterward to give the seedlings a head start, but sometimes just once is enough. Vegetables grown in this way are stronger than most people think. In fact, you can raise produce wherever there's a varied and vigorous growth of weeds ... but to be successful, it is important that you become familiar with the yearly cycle of the indigenous weeds and grasses and learn what kinds of vegetables will best match them.

Mother Earth News: Have you encountered any really serious problems with disease or insect pests over the decades that you've been practicing natural farming?

Fukuoka: Since I turned the fields back to their natural state, I can't say I've had any really difficult problems with insects or disease. Even when it looked as if something had gone wrong and the crops would soon be devastated, nature always seemed to bail me out in the end.

Of course, I have made mistakes . . . just as every grower does. However, I never really think of them as mistakes. Back in the beginning, for example, when 70 percent of a field was overgrown and unproductive and 20 to 30 percent was extremely productive, I saw my limited harvest as a success. I figured that if a small percentage of the field did produce, I could eventually make the rest of the acreage do just as well. My neighbors would never have been satisfied with a field like that . . . but I just viewed the "mistake" as a hint or a lesson. One of the most important discoveries I made in those early years was that to succeed at natural farming, you have to get rid of your expectations. Such "products" of the mind are often incorrect or unrealistic . . . and can lead you to think you've made a mistake if they're not met.

Mother Earth News: What about the wild grasses and weeds that grow right among your crops: Don't they ever threaten to get out of control?

Fukuoka: Instead of relying on herbicides or mechanical cultivation to control weeds, I've always used legumes and other cover crops to limit the spread of the less helpful plants. I also throw straw on the fields as a mulch that will both discourage weeds and let the soil retain enough moisture to germinate seeds in the autumn dry season.

Mother Earth News: It all sounds like the ideal low-labor farming method. But what about the yields of your crops? Is it true that they compare favorably with those of conventional farms?

Fukuoka: In the beginning my expectations and desires were not great . . . and my yields were not great, either. But as the condition of the soil stabilized over time and the fields returned to their natural state, my crop output began to rise steadily. I never noticed any dramatic changes, but eventually I found that I could grow rice without plowing or flooding the field all summer long, and still produce as much as the other farmers did with all their machinery and chemicals . . . sometimes more. My production has now stabilized at about 1,300 pounds, or 22 bushels, per quarter acre for both winter grain and rice. That is close to the highest in Japan!

In the future, I expect that my yields of rice, barley, and other grains will continue to increase. After all, until recently I was growing the same kinds of crops that other farmers in the village—and, indeed, all over Japan—were planting. But as a result of practicing natural agriculture, I have now "developed" some new varieties, simply by allowing them to spring up in the fields. With those native seed cultivars, I think my farm has the potential to achieve the highest productivity in Japan . . . and possibly in the world, since my country leads the planet in average rice yields. If natural farming were used on a permanent basis, there'd be no reason why the production capability of any piece of land couldn't go far beyond its "chemical-based" levels . . . eventually approaching the highest yield theoretically possible, given the amount of energy reaching a field from the sun.

Mother Earth News: I assume that with such favorable production figures you've been able to support yourself and your family using natural farming.

Fukuoka: I haven't made a lot of money, but my overhead costs are so low that I've never been in danger of going completely broke. For one thing, after I began farming this way, word got around that the oranges grown on my mountain were the largest and sweetest in the entire village. That fruit provides the greatest part of my income. Then, too, as my holdings increased and the soil improved, things got easier for us. Yes, I've been able to make a comfortable—though modest—living by practicing natural farming.

Mother Earth News: Has the large-scale agricultural "establishment" exhibited any interest in your ideas?

Fukuoka: I first presented this "direct seeding non-cultivation winter grain/rice succession" plan in agricultural journals 25 years ago. From then on, the method appeared often in print, and I introduced it to the public at large on quite a few radio and television programs . . . but nobody paid much attention to it. In the past 15 years or so, though, it seems to me that the general attitude toward natural farming has begun to change. Various agricultural research scientists have highly acclaimed my no-till technique. You might even say that natural farming is becoming the rage! Journalists, professors, farmers, technical researchers, and students are all flocking to visit my fields and stay in my huts up on the mountain.

By living a natural lifestyle and demonstrating its usefulness, I feel I am serving humankind.

Mother Earth News: Why the sudden surge of curiosity about your farming techniques?

Fukuoka: I think it's because many people have gotten very far away from nature. Everything in the modern world has become noisy and overcomplicated, and people want to return to a simpler, quieter life . . . the kind of life I live as an ordinary farmer. You see, to the extent that men and women separate themselves from nature, they spin out further and further from the unchanging, unmoving center of reality. At the same time a centripetal effect asserts itself, causing a desire to return to nature—that true center—even as they move away from it. I believe that natural farming arises from that unchanging, unmoving center of life.

It seems, also, that general recognition of the long-term dangers of chemical farming has helped renew interest in alternative methods of agriculture. Many people are looking at my methods and seeing that what they previously viewed as primitive and backward is perhaps instead far ahead of modern science.

Mother Earth News: You practice a low-cost, low-labor method of growing food that requires no heavy machinery, fossil fuels, or processed chemicals . . . and yet achieves yields comparable to those of more "modern" scientific methods. That sounds almost like a dream come true. There must be people trying natural farming all over the place.

Fukuoka: Not really . . . because my method does seem like a dream to them. In fact, I think natural farming is actually a very frightening concept to many people. It entails a revolutionary attitude that could change the whole climate of our society and our civilization.

Mother Earth News: What would it take, then, to convince such individuals to try your methods?

Fukuoka: It would be very difficult for single farmers or families to get started by themselves. Natural agriculture requires a great deal of work in the beginning—until the land is brought back into balance—and you can't do it alone unless you have a lot of time to devote to the effort.

The change might be brought about more easily on a village or small-town level, but I really think the best way to start this "one-straw revolution," as I call it, is on a large scale through some sort of cooperative effort. The government, the agricultural co-ops, the farmers, the consumers—in other words, everyone—must decide that this is the direction in which our society should go. And, of course, if we don't get that kind of cooperation, the possibility of bringing about significant change in our farming methods is remote.

Most important, we've got to revise people's concepts of nature. In America, especially, the outdoors that's seen often isn't natural at all . . . it's an imitation, man-made nature. For example, look around the grounds of this university. You'll see beautiful lawns, soft and comfortable, planted here and there with trees. The foliage is indeed lovely, but these aren't the trees and grasses that originally evolved here. They've been put here by human beings for the benefit of other human beings. The native plants were smothered or exterminated . . . and this non-native, exotic lawn grass was nurtured instead. Allowing such an artificial landscape to return to its natural state would be good for human beings and for all the other animals and all the plants that live on this planet. However, not everyone would appreciate it . . . there'd be more flies, mosquitoes, and other insects that people don't find very pleasant, and some would say, "Oh, how inconvenient. What a bother!"

Mother Earth News: Several weeks ago you started your American tour in California. Did you see "artificial nature" there, too?

Fukuoka: It was really a shock for me to see the degenerate condition of California. Ever since the Spanish introduced their grazing cows and sheep, along with such annual pasture grasses as foxtail and wild oats, the native grasses have been all but eliminated. In addition, the groundwater there has been overdrawn for agriculture, and huge dams and irrigation projects have interrupted the natural circulation of surface water. Forests have been logged heavily and carelessly, causing soil erosion and damage to streams and fish populations. As a result of all this, the land is becoming more and more arid. It's a dreadful situation . . . because of human intervention, the desert is creeping across the state, but no one will admit it.

Mother Earth News: Do you think the widespread adoption of natural farming techniques could help reverse that process and make California green again?

Fukuoka: Well, it would take a few years for people to learn how to adjust and refine the weed/ground cover rotation, but I think the soil would improve rapidly if growers really attempted to help it. And if that were done, California could eventually become an exciting, truly natural place . . . where farming could be the joyous activity it should be. But if modern agriculture continues to follow the path it's on now, it's finished. The food-growing situation may seem to be in good shape today, but that's just an illusion based on the current availability of petroleum fuels. All the wheat, corn, and other crops that are produced on big American farms may be alive and growing, but they're not products of real nature or real agriculture. They're manufactured rather than grown. The earth isn't producing those things . . . petroleum is!

Mother Earth News: Haven't you said that you'd view a severe oil shortage as a positive development?

Fukuoka: Of course. I believe that the sooner our oil supply lines dry up, the better. Then we'll have no choice but to turn to natural agriculture.

Mother Earth News: But the typical agribusiness farm has hundreds or even thousands of cultivated acres. How could someone apply natural agriculture in such a setting?

Fukuoka: First of all, there shouldn't be such large spreads. It's unfortunate that, in the modern American agricultural system, a very few people are producing the food for millions of others who live in the cities. In Japan, the average field is smaller than in the United States ... but its yield per acre is much greater. I can do all the work on my own farm with hand tools, without using power machinery of any kind.

But I guess those mega-farms in your country would need some machinery, at least for harvesting. In the future, though, as more and more people move back to the country and begin to grow their own food on small plots of land, there'll be much less dependence on machines and fossil fuels . . . and natural farming techniques can begin to be used.

Mother Earth News: So you think that it would be feasible to someday adopt natural farming in North America?

Fukuoka: Of course, of course! When you talk about nature, it doesn't matter whether you're referring to North America or Africa or Indonesia or China ... nature is nature. After all, modern industrial farming is now being practiced almost everywhere in the world. In the same way, natural farming could be practiced almost everywhere.

I'm just a village farmer who has come visiting from another part of the same world. Through my one-straw research, I've come up with some important clues as to how people can relate to nature and live harmoniously with it ... wherever they may be.

Mother Earth News: Some people have noticed a spiritual, almost mystical quality to your theory of farming. Do you feel you're receiving insight and guidance from a divine source?

Fukuoka: Although natural farming—since it can teach people to cultivate a deep understanding of nature—may lead to spiritual insight, it's not strictly a spiritual practice. Natural farming is just farming, nothing more. You don't have to be a spiritually oriented person to practice my methods. Anyone who can approach these concepts with a clear, open mind will be starting off well. In fact, the person who can most easily take up natural agriculture is the one who doesn't have any of the common adult obstructing blocks of desire, philosophy, or religion ... the person who has the mind and heart of a child. One must simply know nature . . . real nature, not the one we think we know!

Mother Earth News: Can you be more specific about what that attitude should be?

Fukuoka: Many people think that, when we practice agriculture, nature is helping us in our efforts to grow food. That is an exclusively human-centered viewpoint . . . we should, instead, realize that we are receiving that which nature decides to give us. A farmer does not grow something in the sense that he or she creates it. That human is only a small part of the whole process by which nature expresses its being. The farmer has very little influence over that process . . . other than being there and doing his or her small part.

People should relate to nature as birds do. Birds don't run around carefully preparing fields, planting seeds, and harvesting food. They don't create anything . . . they just receive what is there for them with a humble and grateful heart. We, too, receive our nourishment from the Mother Earth. So we should put our hands together in an attitude of prayer and say "please" and "thank you" when dealing with nature.

Mother Earth News: Do you think that, partly by helping foster such different attitudes, your method could influence more than the way we grow our food?

Fukuoka: Yes, natural farming could lead to changes in our way of life that would help solve many of the problems of our present age. I think that people are starting to have misgivings about the modern world's ever-accelerating growth and scientific development, to question such things as nuclear power plants and the massive slaughter of great whales, and to realize that the time for reappraisal has arrived.

By living a natural lifestyle and demonstrating its usefulness in this day and age, I feel I am serving humankind. As the steward of my rice fields, I am making my stand against the need to use destructive technology or eliminate other forms of life. After all, the problems of our time are ones all of us must face in our own hearts and deeds. As I see it, the ultimate goal of natural farming is not the growing of crops . . . but the cultivation and perfection of human beings.

APPENDIX D

Making Clay Seed Pellets for Use in Revegetation (from *Sowing Seeds in the Desert*)[1]

Purpose

The clay seed pellet was conceived and developed for direct seeding rice, barley, and vegetables in conjunction with the no-till method. It has since come into wide use, and is particularly well suited for aerial seeding for the purpose of revegetating large areas of desert at one time.

Materials

1. Seeds of more than one hundred varieties (trees, fruit trees, shrubs, vegetables, grains, useful fungi). Ten percent of combined weight.*
2. Fine powdered clay such as that used for fired bricks or porcelain. In general, this should make up five times the weight of the seeds, but the amount of seeds should be taken into consideration. Fifty percent of the combined weight.
3. Bittern—the liquid remaining after removing salt from the brine obtained by boiling and concentrating seawater or from natural brackish water (such as the water found in the Dead Sea). Ten to 15 percent of combined weight, with seaweed paste for binding making up 5 percent of combined weight.

---

* In later years and after many trials in different parts of the world, Mr. Fukuoka revised the ratio of seeds to powdered clay from 1:5, as given here, to 1:20, in other words, far fewer seeds for each pellet.

4. Slaked lime—10 percent of combined weight.
5. Medicinal herbs: derris (root), powdered fruits and leaves of Japanese star anise (*Illicium anisatum*), Japanese andromeda (*Pieris* spp.), Japanese lacquer tree (*Rhus verniciflua*), Japanese bead tree (*Melia azedarach*). Ten percent of combined weight.
6. Water—5 to 10 percent of combined weight.

Aerial Seeding Method (Overseas)

The seeds necessary for revegetation of the desert will be mixed in clay pellets and broadcast from airplanes or by hand to revegetate large areas at one stroke.

Method of Production

When producing pellets in large quantities, a typical concrete mixer (with inner blades removed) is useful.

1. Put fungi and seeds into the mixer and mix well to spread the fungi about (inner layer).
2. Next, alternately add the clay powder with a water mist in the mixer as it is rotating, to create a layer enclosing the seeds and fungi (middle layer).
3. Then, when you alternately add and spray the bittern, seaweed paste solution, clay powder, and slaked lime into the mixer as it is rotating, a round seed pellet usually about a quarter inch to a half inch in diameter will form (outer layer).

Properties

1. The seeds enclosed in the layers of clay will achieve satisfactory germination and growth with the aid of the useful fungi.
2. By kneading the clay together with the bittern and seaweed paste, its molecules are rearranged, so the pellets become stable, light, and hard. They will not only withstand the fall to earth following aerial seeding, but also adjust to changes in dampness and dryness related to rainfall, becoming shrunken and solid. Thus, they seldom crumble or break, and the seeds are protected from damage by birds or animals until they germinate.

3. Many insects are repelled by the bitterness of the herbs and the bittern mixed into the outer layer, so the seeds escape being eaten. In deserts and savannas this technique helps prevent damage by mice, goats, and in particular strong insects such as red ants. Even damage by birds can be prevented simply by enclosing the seeds in pellets. The method described here not only ensures safe germination of seeds in desert areas without the use of toxic substances, but also makes it possible to indiscriminately broadcast the seeds over a wide area.
4. The plants on earth exist in intimate connection with other plants, animals, and microorganisms, and none can develop and flourish alone. In desert regions, in particular, microorganisms are necessary as well as a variety of plants.
5. Derris root (used against beetles), Japanese star anise (goats), Japanese andromeda (cows), Japanese bead tree (small insects), sumac, and so on, will protect seeds in the desert, before and after germination.

In a region that is completely desert, it is a good idea to mix fertile jungle or forest soil with the clay. This soil is a rich source of soil microorganisms, seeds, and spores, and is of great value when added to the pellets.

If the pellets will be broadcast from airplanes, the pellets may break on contact with the ground, so it is good to coat them with seaweed paste, if available.

Thus, even in vast desert areas, where conditions for germination are poor, revegetation can be achieved simply by sowing the seeds, without concern for time or place. Successful results have already been achieved in Africa, the United States, India, Greece, and the Philippines.

Bibliography

Anderson, M. Kat. *Tending the Wild: Native American Knowledge and the Management of California's Natural Resources*. Berkeley, Los Angeles: University of California Press, 2005.

Beardsley, Richard K., et al. *Village Japan*. Chicago, London: University of Chicago Press, 1959.

Berry, Wendell. *The Unsettling of America: Culture and Agriculture*. San Francisco: Sierra Club Books, 1977.

Blair, Katrina. *The Wild Wisdom of Weeds: 13 Essential Plants for Human Survival*. White River Junction, VT: Chelsea Green Publishing, 2014.

Brown, Azby. *Just Enough: Lessons in Living Green from Traditional Japan*. Tokyo, New York, London: Kodansha International, 2009.

Burr, Chuck. *Culturequake*. 3rd edition. Bloomington, IN: Trafford Publishing, 2010.

Faulkner, Edward H. *Plowman's Folly*. New York: Grosset & Dunlap (by arrangement with University of Oklahoma Press), 1943.

Fukuoka, Masanobu. *The One-Straw Revolution: An Introduction to Natural Farming*. Emmaus, PA: Rodale Press, 1978.

———. *The Natural Way of Farming: The Theory and Practice of Green Philosophy*. Tokyo, New York: Japan Publications, 1985.

———. *The Road Back to Nature: Regaining the Paradise Lost*. Tokyo, New York: Japan Publications, 1987.

———. *Sowing Seeds in the Desert: Natural Farming, Global Restoration, and Ultimate Food Security*. White River Junction, VT: Chelsea Green Publishing, 2012.

Heizer, Robert F., and Albert B. Elasser. *The Natural World of the California Indians*. Berkeley, Los Angeles: University of California Press, 1980.

Hemenway, Toby. *Gaia's Garden: A Guide to Home-Scale Permaculture*. 2nd edition. White River Junction, VT: Chelsea Green Publishing, 2009.

Howard, Sir Albert. *An Agricultural Testament*. London, New York, Toronto: Oxford University Press, 1943.

———. *The Soil and Health: A Study of Organic Agriculture*. New York: Devin-Adair, 1947.

Jacke, Dave, with Eric Toensmeier. *Edible Forest Gardens*. Vols. 1 and 2. White River Junction, VT: Chelsea Green Publishing, 2005.

King, F. H. *Farmers of Forty Centuries or Permanent Agriculture in China, Korea and Japan*. Reprint from original plates. Emmaus, PA: Rodale Press, 1911.

Korn, Larry, et al., eds. *The Future Is Abundant: A Guide to Sustainable Agriculture*. Arlington, WA: Tilth, 1982.

Mansata, Bharat. *The Vision of Natural Farming*. Kolkata, India: Earthcare Books, 2010.

Margolin, Malcolm. *The Ohlone Way: Indian Life in the San Francisco–Monterey Bay Area*. Berkeley, CA: Heyday Books, 1978.

Quinn, Daniel. *Ishmael*. New York: Bantam Books, 1992.

———. *The Story of B: An Adventure of the Mind and Spirit*. New York: Bantam Books, 1997.

Roberts, Harry K. *Walking in Beauty: Growing Up with the Yurok Indians*. Trinidad, CA: The Press at Trinidad Art, 2011.

Rodale, J. I. *Pay Dirt: Farming and Gardening with Composts*. New York: Devin-Adair, 1946. (Introduction by Sir Albert Howard.)

Rush, James R. Interview from tape recording. Manila, Philippines: Ramon Magsaysay Award Foundation, September 1988.

Sahlins, Marshall. *Stone Age Economics*. New York: Aldine de Gruyter, 1972.

Sauer, Carl Ortwin. *Land and Life: A Selection from the Writings of Carl Ortwin Sauer*. Edited by John Leighly. Berkeley, Los Angeles: University of California Press, 1963.

Smith, Russell J. *Tree Crops: A Permanent Agriculture*. New York: Devin-Adair, 1953.

White, Courtney. *Grass, Soil, Hope: A Journey Through Carbon Country*. White River Junction, VT: Chelsea Green Publishing, 2014.

Notes

Introduction

1. Fukuoka, 1978, page 118.
2. Ibid., page 159.

Chapter 1

1. Rush, 1988.
2. Fukuoka, 1978, pages 8–9.
3. Rush, 1988.
4. Fukuoka, 1978, page 15.
5. Fukuoka, 1978, page 119.
6. Leath Tonino, "The Egret Lifting from the River: David Hinton on the Wisdom of Ancient Chinese Poets" (interview with David Hinton), *The Sun* 469 (January 2015), pages 6–9.
7. Ibid., page 13.
8. Berry, 1977, page 26.
9. Fukuoka, 2012, page 22.
10. Ibid.
11. Fukuoka, 1978, page 17.

Chapter 2

1. Damari, "An Introduction to Suwanose Ontake, Fire Island," *Om* (a periodical of the Cosmic Child Community, Tokyo), December 1974.
2. Ibid.

Chapter 3

1. Fukuoka, 1985, page 120.

2. Fukuoka, 1978, pages 143, 136.

Chapter 4

1. Fukuoka, 1978, page 126.
2. T. F. Cronise, *The Natural Wealth of California* (San Francisco: H. H. Bancroft, 1868) (as quoted in Anderson, 2005, page 13).
3. Fukuoka, 1987, page 56.

Chapter 5

1. firstpeoples.org/how-our-society-works. Accessed March 15, 2015.
2. Fukuoka, 1978, pages 127, 138.
3. B. Callahan, ed. *A Jaime de Angulo Reader* (Berkeley, CA: Turtle Island, 1979), page 240 (as quoted in Anderson, 2005, page 58).
4. Anderson, 2005, page 57.
5. Ibid., page 58.
6. S. Powers, *Tribes of California* (Berkeley: University of California Press, 1976), pages 109–110 (as quoted in Anderson, 2005, page 39).
7. Anderson, 2005, page 55.
8. Ibid
9. Heizer and Elsasser, 1980, page 27.
10. *Akwesasne Notes* 10, no. 5 (winter 1978), pages 30–32.
11. Anderson, 2005, page 3.

Chapter 6

1. Japan for Sustainability Staff, "Japan's Sustainable Society in the Edo Period (1603–1867)," *Japan for Sustainability Newsletter*, April 5, 2005.
2. Brown, 2009, page 48.
3. Ibid., page 10
4. King, 1911, pages 9, 11.
5. Ibid., page 401.
6. Brown, 2009, page 63.
7. Fukuoka, 1985, page 30.
8. Ibid., page 31.

Chapter 7

1. Berry, 1977, pages 7–8.

2. Albert Howard, from the introduction to Rodale, 1946.
3. King, 1911, pages 274, 276.
4. B. F. Lutman, "The Scientific Work of Sir Albert Howard," *Organic Gardening Magazine* 13, no. 8 (September 1948).
5. Louise Howard, "The Work at Indore," *Organic Gardening Magazine* 13, no. 8 (September 1948).
6. Howard, 1947, page 35.
7. Ibid.
8. Ibid., page 33.
9. Ibid.
10. Faulkner, 1943, pages 4–5.
11. Russell Lord, "Two Revolutions in Plowing," *The Nation*, October 9, 1943, pages 412–13.
12. Smith, 1953, page 3.
13. Ibid., page 8.
14. Ibid., pages 317–18.
15. Fukuoka, 1978, page 109.
16. Jone Johnson Lewis, "Ruth Stout Quotes," *About Education*, http://womenshistory.about.com/od/quotes/a/ruth_stout.htm, accessed August 31, 2013.

Chapter 8

1. Blair, 2014, pages 29–31.

Appendix A

1. Roberts, 2011, pages 56–58.

Appendix B

1. *Akwesasne Notes*, 1978, pages 30–32.

Appendix C

1. Interview with Masanobu Fukuoka, *Mother Earth News* 76 (July–August 1982).

Appendix D

1. From Fukuoka, 2012, pages 161–65.

Index

Bold entries refer to the color insert.

About the Author

Larry Korn is an American who lived and worked on the farm of Masanobu Fukuoka for more than two years in the early 1970s. He is translator and editor of the English-language edition of Mr. Fukuoka's *The One-Straw Revolution* and editor of his later book, *Sowing Seeds in the Desert*. Korn accompanied Mr. Fukuoka on his visits to the United States in 1979 and 1986. He studied Asian history, soil science, and plant nutrition at the University of California, Berkeley, and has worked in wholesale and retail plant nurseries, as a soil scientist for the California Department of Forestry, and as a residential landscape contractor in the San Francisco Bay Area. Korn has taught many courses and workshops about natural farming, permaculture, and local food production throughout the United States. He currently lives in Ashland, Oregon.

Chelsea Green Publishing is committed to preserving ancient forests and natural resources. We elected to print this title on paper containing 100% postconsumer recycled paper, processed chlorine-free. As a result, for this printing, we have saved:

38 Trees (40' tall and 6-8" diameter)
17,688 Gallons of Wastewater
17 million BTUs Total Energy
1,185 Pounds of Solid Waste
3,261 Pounds of Greenhouse Gases

Chelsea Green Publishing made this paper choice because we are a member of the Green Press Initiative, a nonprofit program dedicated to supporting authors, publishers, and suppliers in their efforts to reduce their use of fiber obtained from endangered forests. For more information, visit www.greenpressinitiative.org.

Environmental impact estimates were made using the Environmental Defense Paper Calculator. For more information visit: www.papercalculator.org.